FM 17-80

DEPARTMENT OF THE ARMY TECHNICAL MANUAL

TANKS, 76-MM GUN M41 AND M41A1 WALKER BULLDOG TECHNICAL MANUAL

By DEPARTMENT OF THE ARMY • JANUARY 1956

AGO 4054B—Jan

©2013 Periscope Film LLC
All Rights Reserved
ISBN#978-1-940453-08-8
www.PeriscopeFilm.com

DISCLAIMER:

This document is a reproduction of a text first published by the Department of the Army, Washington DC. All source material contained herein has been approved for public release and unlimited distribution by an agency of the US Government. Any US Government markings in this reproduction that indicate limited distribution or classified material have been superseded by downgrading instructions promulgated by an agency of the US government after the original publication of the document No US government agency is associated with the publication of this reproduction. This manual is sold for historic research purposes only, as an entertainment. It contains obsolete information and is not intended to be used as part of an actual training program. No book can substitute for proper training by an authorized instructor.

©2013 Periscope Film LLC
All Rights Reserved
ISBN#978-1-940453-08-8
www.PeriscopeFilm.com

FIELD MANUAL
No. 17-80

DEPARTMENT OF THE ARMY
WASHINGTON 25, D. C., *30 January 1956*

TANKS, 76-mm GUN, M41 AND M41A1

		Paragraphs	Page
CHAPTER 1.	INTRODUCTION	1	3
2.	MATERIEL		
Section I.	General characteristics of the tanks	2-4	4
II.	Gun, 76-mm, M32	5-13	8
III.	Machine gun mounts	14-18	29
IV.	Turret and armament controls and equipment, M41 tank.	19-26	33
V.	Turret and armament controls and equipment, M41A1 tank.	27-34	42
VI.	Sighting and fire-control equipment	35-51	48
CHAPTER 3.	CREW DRILL AND SERVICE OF THE PIECE		
Section I.	General	52	72
II.	Crew compositions and formations	53, 54	72
III.	Crew control	55-61	72
IV.	Crew drill	62-67	76
V.	Service of the piece	68-75	80
VI.	Mounted action	76-81	83
VII.	Dismounted action	82-84	90
VIII.	Evacuation of wounded from tanks	85-89	91
IX.	Destruction of equipment	90	94
CHAPTER 4.	CREW PREVENTIVE MAINTENANCE		
Section I.	Introduction	91, 92	95
II.	Crew maintenance procedures	93-97	96
CHAPTER 5.	CONDUCT OF FIRE		
Section I.	Introduction	98	104
II.	Firing duties	99	104
III.	Firing at stationary targets	100-104	105
IV.	Firing at moving targets	105-111	114
V.	Firing tank machine guns	112	122
VI.	Special situations	113-121	123
CHAPTER 6.	TANK GUNNERY QUALIFICATION COURSE		
Section I.	Introduction	122-126	128
II.	Gunner's preliminary examination, general	127, 128	130

* This manual supersedes so much of chapter 18, FM 17-12, 30 November 1950, as pertains to Tanks, 76-mm Gun, M41 and M41A1.

		Paragraphs	*Page*
Section III.	Materiel tests, gunner's preliminary examination.	129–136	131
IV.	Simulated firing tests, gunner's preliminary examination.	137–141	133
V.	Subcaliber firing exercises	142–146	137
VI.	Service firing exercises	147–151	149
APPENDIX I.	REFERENCES		159
II.	STOWAGE		160
INDEX			172

CHAPTER 1

INTRODUCTION

1. Purpose and Scope

   a. The purpose of this manual is—
   (1) To give the general characteristics of the Tanks, 76-mm Gun, M41 and M41A1.
   (2) To explain, in detail, the 76-mm Gun, M32, and the turret controls, fire-control instruments, and auxiliary fire-control instruments in the Tanks, 76-mm Gun, M41 and M41A1.
   (3) To provide a guide for Armor personnel in learning and teaching fire commands, firing duties, crew drill, and service of the piece for the M41 and M41A1 tanks.
   (4) To describe the procedures for conduct of fire.
   (5) To outline the Tank Gunnery Qualification Course, consisting of the Gunner's Preliminary Examination and the firing courses.

   b. This manual covers materiel, crew drill, service of the piece, conduct of fire, and the Tank Gunnery Qualification Course for the Tanks, 76-mm Gun, M41 and M41A1.

CHAPTER 2
MATERIEL

Section I. GENERAL CHARACTERISTICS OF THE TANKS

2. General

The Tanks, 76-mm Gun, M41 and M41A1, resulted from several years of research and development on a high-speed tank for use in reconnaissance and security missions, capable of being transported by air. First models of the tank came off the production line in mid-1951. Basic rudiments of the hull, turret, and suspension system were originally developed in the M24 light tank during World War II. The M41 and M41A1, however, incorporate many refinements and improvements resulting from subsequent research and development.

3. Description, Tanks, 76-mm Gun, M41 and M41A1

a. The Tanks, 76-mm Gun, M41 and M41A1 (figs. 1, 2, 3, and 4), are lightly armored, full-track, low-silhouette, combat vehicles, each manned by four crewmen and armed with a 76-mm high-velocity, flat-trajectory gun and a caliber .30 coaxial machinegun. A caliber .50 machinegun is mounted on the turret for use by the tank commander. Each tank has a forward hull compartment for the driver and for ammunition stowage, a rear compartment for the engine and transmission, and a turret mounting the main armament and providing stations for the tank commander, gunner, and loader.

b. The M41 and M41A1 (which is an improved model of the M41) differ in that the M41 has a power pack mounted on the turret floor and a traversing mechanism mounted in the right front of the turret, and has a manual hydraulic elevation system; whereas the M41A1 power pack and traversing mechanism are combined in a single self-contained unit mounted in the right front of the turret, and there is a mechanical elevation system on the right of the gun. In addition, the M41A1 provides the tank commander with means for constant-speed power elevation of the tank gun. These differences are described in detail in paragraphs 19 through 34.

Figure 1. Tank, 76-mm Gun, M41 or M41A1, right side (gun in firing position).

AGO 4054B

Figure 2. Tank, 76-mm Gun, M41 or M41A1, left side (gun in travel lock).

Figure 3. Tank, 76-mm Gun, M41 or M41A1, rear (gun in travel lock).

4. Data, Tanks, 76-mm Gun, M41 and M41A1

Crew	Four.
Armament	One Gun, 76-mm, M32; one caliber .30 coaxial machinegun; one caliber .50 turret-mounted machinegun.
Communication system	Radio and interphone.
Weight, full equipped (approx.)	26 tons.
Length, overall (gun in traveling position)	22 feet, 11 inches.
Length, overall (gun in firing position)	26 feet, 5 inches.
Width	10 feet, 8 inches.
Height (overall)	9 feet.
Ground clearance	17¼ inches.
Ground pressure	9.7 pounds per square inch.
Engine (Continental 6-cylinder, gasoline, air cooled)	500 hp @ 2,800 rpm.
Electrical system	24 volts.
Maximum allowable speed	40 mph.
Maximum grade-ascending ability	60 percent.

Figure 4. Tank, 76-mm Gun, M41 or M41A1, front (gun in firing position).

```
Maximum trench-crossing ability_____6 feet.
Maximum vertical obstacle_____2 feet, 4 inches.
Maximum fording depth_____3 feet, 4 inches.
Maximum turning circle (stationary)___Pivot.
```

Section II. GUN, 76-MM, M32

5. General

  a. The Gun, 76-mm, M32 (fig. 5), is designed for the light-gun tank. It consists of four major parts: the tube, bore evacuator, blast deflector (or muzzle brake), and breech mechanism.

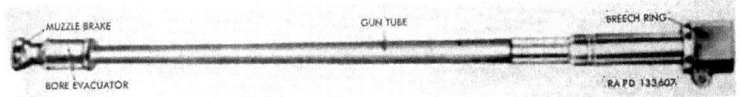

Figure 5. Gun, 76-mm, M32 (less mount).

b. The tube is formed in one piece, threaded at the breech end for attachment of the breech mechanism and at the muzzle end for assembly of the bore evacuator and the blast deflector. The bore of the tube is rifled with grooves having a uniform right-hand twist, and the tube is chrome-plated over the recoil slide surface.

c. The bore evacuator is formed by a thin-walled cylinder fitted over the front end of the tube to form an evacuator chamber. Seven holes, drilled into the tube and slanted at an angle of 30° toward the muzzle, connect the evacuator chamber to the bore. The bore evacuator removes residual gases from the gun tube through the muzzle end, preventing crew discomfort caused by these gases escaping into the fighting compartment.

d. The blast deflector is a heavy sleeve threaded at one end for attachment to the gun tube. It has a baffle on the other end to deflect muzzle blast to the side. It is designed to balance the gun and to reduce obscuration of the target by muzzle blast.

e. The breech mechanism consists of the breech ring, breechblock with its component parts, and breech operating mechanism.

6. Data, Gun, 76-mm, M32

Caliber	76 millimeters.
Length of bore	15 feet.
Type of breechblock	Vertical sliding block.
Maximum chamber pressure	46,000 pound per square inch.
Type of recoil mechanism	Concentric, hydrospring.
Normal recoil	9 inches.
Maximum recoil	12 inches.
Maximum elevation	19¾° (351 mils).
Maximum depression	9¾° (173 mils).
Traverse	360° (6400 mils).
Weight of gun complete	1,320 pounds.

Ammunition and approximate velocity

AP–T	3,200 feet per second.
HEAT	2,800 feet per second.
HVAPDS–T	4,125 feet per second.
HE–T	2,400 feet per second.
Smoke (WP–T)	2,400 feet per second.

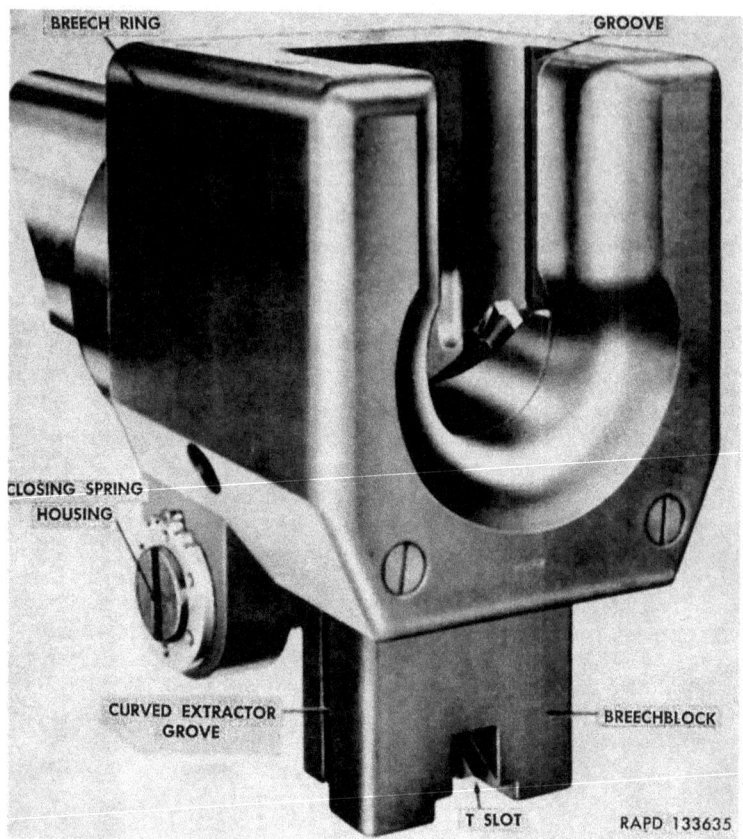

Figure 6. Breech mechanism (left side).

7. Breech Mechanism, Gun, 76-mm, M32

a. Breech Operation. The breech operation of the 76-mm Gun, M32, is semiautomatic after the breechblock has been moved to the open position. When a round of ammunition is loaded, the rim of the case trips the extractors, which were holding the breechblock in the open position. A closing spring then functions to move the breechblock to the closed position, fully seating the round in the chamber. The gun is fired by means of an inertia-type percussion mechanism located in the breechblock. The breechblock remains closed as the gun recoils. During counterrecoil, an operating cam and operating crank function to move the breechblock to the open position. As the breechblock nears the open position, the extractors are rotated to extract the empty cartridge case from the chamber,

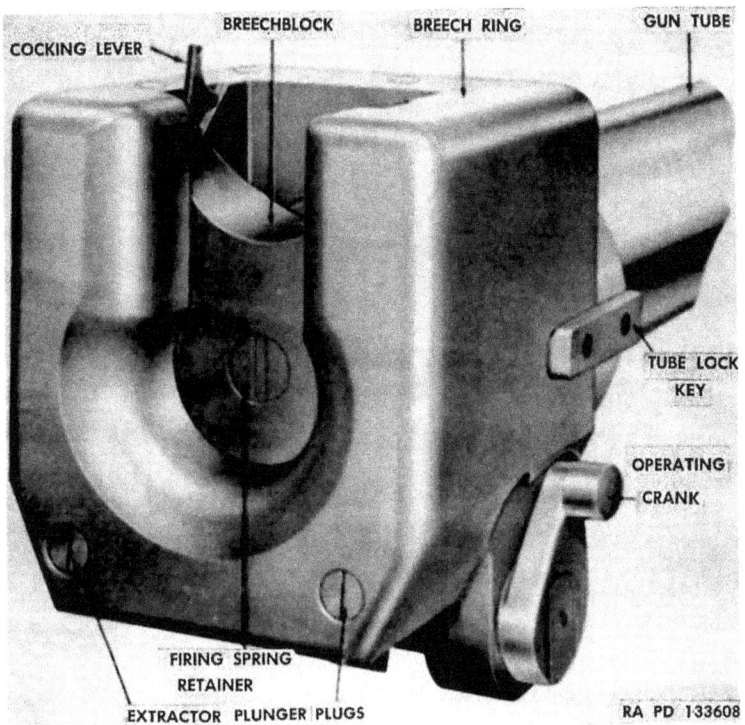

Figure 7. Breech mechanism (right side).

eject it clear of the breech ring, and lock the breechblock in the open position.

  b. *Breech Mechanism.* The breech mechanism (figs. 6, 7, and 8) consists of the breech ring, breechblock and its component parts, and breech operating mechanism.

> (1) *Breech ring.* The rectangular portion of the breech ring (fig. 9) has a central vertical opening, machined to accommodate the operation of the breechblock. The loading opening in the rear, having parallel sides and a concave bottom, also provides an opening for the ejection of the shell casing. The two lugs on the bottom retain the operating crank and the closing spring housing. The breechblock crank stop is retained in a slot at the bottom. Three quadrant seats are inlaid in the top surface. Two inner vertical walls are recessed to retain the extractors.
>
> (2) *Breechblock.* The breechblock (fig. 10) is a vertical operating block. The top is concave to permit inserting a round in the chamber when the breech mechanism is in

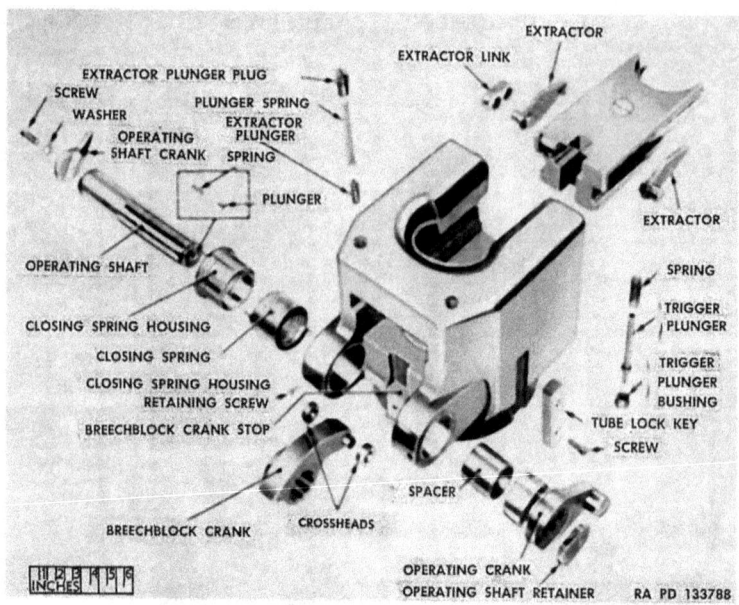

Figure 8. Breech mechanism (exploded view).

the open position. The front top edge is beveled to force the round into the chamber as the breech closes. The bottom of the breechblock contains an inclined T-slot in which the crossheads on the breechblock crank operate to open or close the breech. Each side of the breechblock contains an extractor groove through which the extractor trunnions operate. This groove is curved to the front to impart a quick forward thrust to the extractor trunnions, with a resultant rearward thrust of the extractor lips to extract and eject the empty shell casing. The percussion mechanism is located in the center rear face of the breechblock. The trigger, sear, and sear spring are located in the right side; and the cocking lever, cocking bar, cocking bar pivot, pivot plunger spring, and plug are in the left side.

(3) *Extractors.* The extractors, right and left, are supported vertically on the inside walls of the breech ring. The lower end of each extractor (as installed) has an inner trunnion with a flat surface operating in the breechblock groove and an outer trunnion supported by the extractor link. The lip at the top of each extractor fits ahead of the shell casing rim when the breech is closed.

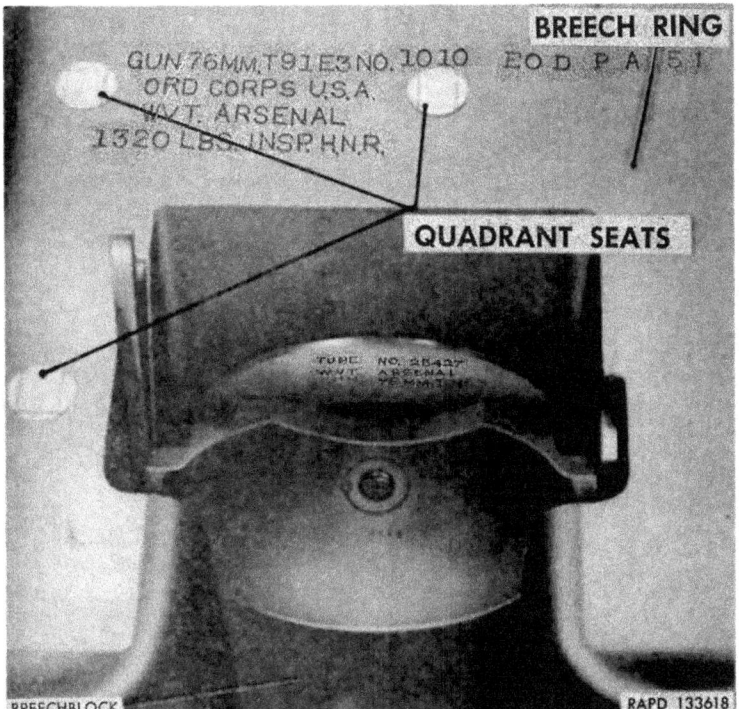

Figure 9. Breech ring, showing quadrant seats.

(4) *Breech operating mechanism* (fig. 8). The breech operating mechanism consists of the operating shaft, operating shaft crank, breechblock crank, operating crank, closing spring, closing spring housing, extractors, extractor links, and operating handle.

  (*a*) The operating shaft is externally splined to mate with the internal spline of the operating crank and breechblock crank. This shaft is supported by the operating crank and breech ring lug on the right, and the closing spring housing and breech ring lug on the left. The operating shaft retainer is secured to the operating shaft with a plunger and spring to prevent side movement of the shaft.

  (*b*) The breechblock crank (fig. 8) connects the breechblock to the operating shaft. The hub is internally splined to fit the operating shaft. The upper curved arm of the breechblock crank has two pivots on which

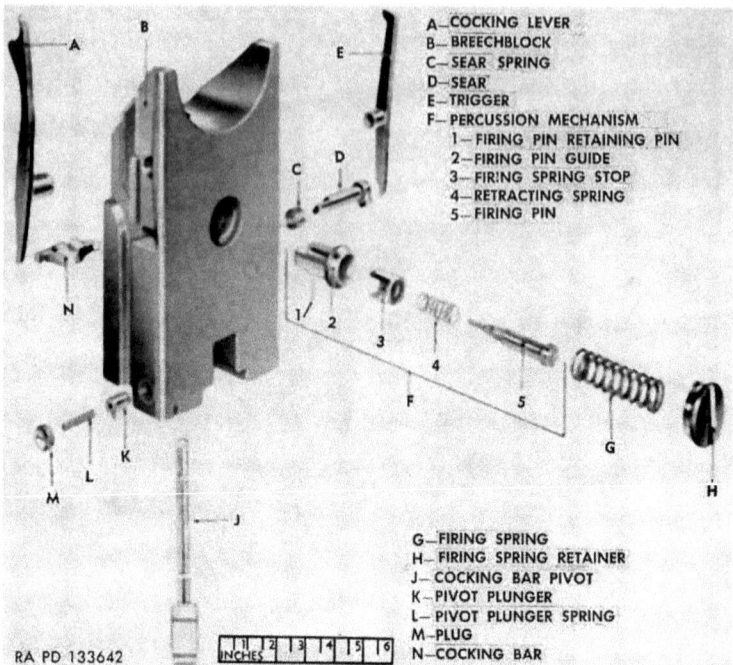

Figure 10. Breechblock (exploded view).

the crossheads are placed. The crossheads fit into, and move in, the inclined T-slot of the breechblock. The crank is mounted in the center of the operating shaft between the spacer and the closing spring. The lower shoulder, or straight face of the crank, contacts the breechblock crank stop on the bottom of the breech ring when the breech is opened.

(c) The operating crank (fig. 8) is a short sleeve with a collar, and an offset lug on the end. The inside of the crank is splined to fit the operating shaft and is held in the lower right lug of the breech ring by the operating crank retaining screw.

(d) The operating handle (fig. 11) is secured with a cap screw to a crank that has been inserted through a machined hole in the left breech guard. The crank contacts the operating shaft crank when the breech is operated manually.

(e) The closing mechanism (fig. 12) is composed of the closing spring, closing spring housing, and closing

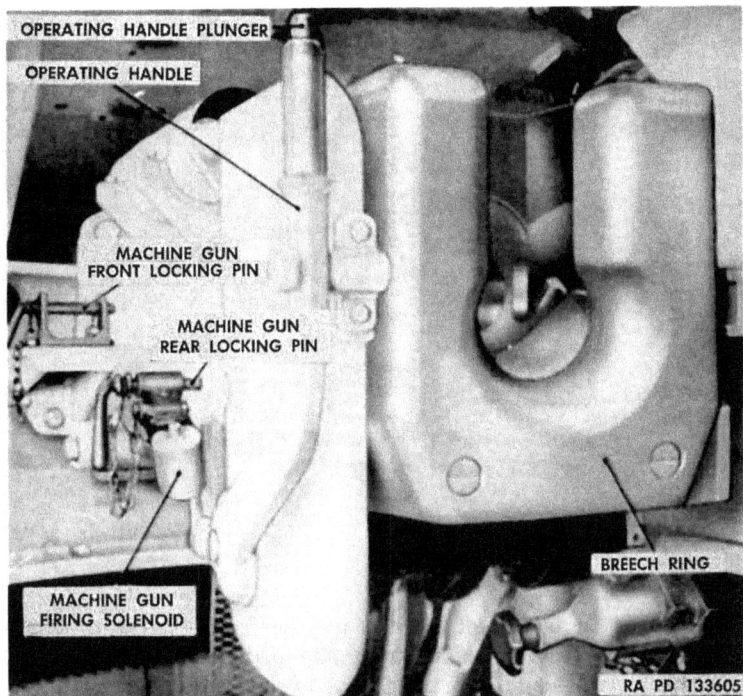

Figure 11. Breech ring and left recoil guard.

spring housing plunger and spring. The closing spring housing is held in the left breech ring lug by a screw and can be rotated to adjust tension on the closing spring. The adjustment is retained by the closing spring housing plunger and spring, which are located in the left side of the breech ring and engage in one of several notches on the housing flange. The left end of the closing spring is anchored to the closing spring housing. The right end of the closing spring fits into a notch in the breechblock crank and exerts a rotational force directly on the crank. The operating shaft is positioned through the closing spring housing, closing spring, breechblock crank, spacer, and operating crank.

8. Cocking and Firing Mechanisms, Gun, 76-mm, M32

   a. Cocking Parts. The cocking lever (fig. 10) is flat, with a long upper arm and a short lower arm. The cocking lever pivots on a hub which fits into a recess in the upper left side of the breechblock. The upper arm has a curved lug projecting forward to

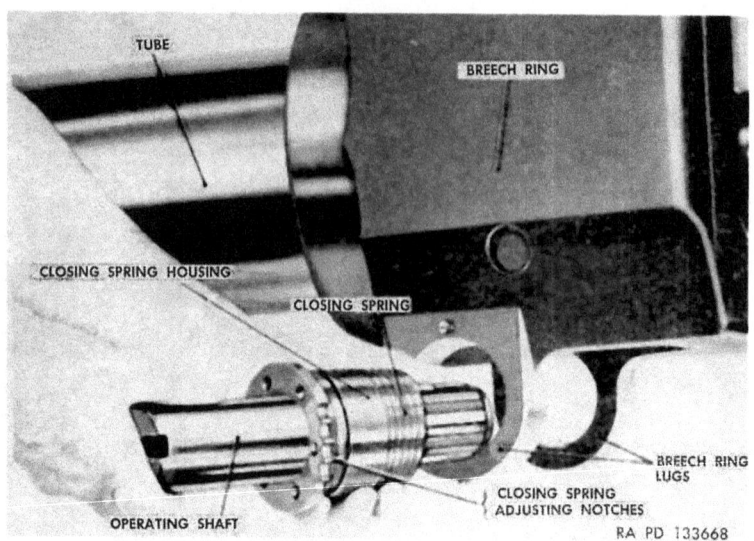

Figure 12. Closing mechanism.

obtain pivoting action of the cocking lever from the camming surface of the breech ring. The upper arm projects above the breechblock to permit manual cocking. The lower arm contacts the outer arm of the cocking bar (fig. 10), which pivots on the cocking bar pivot. The inner arm of the cocking bar engages the collar of the percussion mechanism, moving it to the rear in cocking. The cocking bar and pivot are repositioned as the breech closes by the cocking bar pivot plunger and spring (fig. 10), located in the lower portion of the breechblock.

b. Firing Parts. The firing parts (fig. 10) include the percussion mechanism, firing spring, firing spring retainer, sear, sear spring, and trigger.

    (1) *Percussion mechanism.* The percussion mechanism (fig. 10) is an inertia-type firing device composed of the firing pin guide, firing pin, firing pin retaining pin, retracting spring, and firing spring stop.

        (*a*) The guide is a cylindrical cup which slides forward and backward in the percussion mechanism well of the breechblock with the collar to the rear. The guide houses the firing pin and its retaining pin, retracting spring, and firing spring stop.

        (*b*) The firing pin is a shouldered screw with a small cylindrical point at the front which strikes the ammunition primer.

(c) The retracting spring fits over the firing pin and is held in position between the head of the firing pin and the firing spring stop.

(d) The firing spring stop is a small cylinder with two lugs protruding in front that fit into corresponding holes in the front of the firing pin guide. The firing pin moves through the firing spring stop to contact the primer when the gun is fired.

(2) *Firing spring.* The firing spring (fig. 10) is a compression-type spring. The forward part of the spring fits into the guide over the firing pin and is compressed between the firing spring stop at the front and the firing spring retainer at the rear.

(3) *Firing spring retainer.* The firing spring retainer (fig. 10) is a flat cup with two projecting lugs. These lugs fit into corresponding grooves in the rear of the percussion mechanism well. The rear face of the retainer is slotted to expedite disassembly and assembly.

(4) *Sear.* The sear is a rod which fits into the sear well in the right side of the breechblock (fig. 10). The left end is cut to form a lip which engages the collar of the percussion mechanism, holding it in the cocked position. The right end of the sear has an offset stud which is contacted by the lower arm of the trigger. The sear spring fits around the sear.

(5) *Trigger.* The trigger (fig. 10) is a tapered, flat lever with a hub in the center which fits into a hole in the breechblock and is the trigger pivot. The trigger is mounted in a groove in the right side of the breechblock.

(6) *Trigger plunger.* The trigger plunger (fig. 8) is a cylindrical rod, mounted in a hole in the upper right corner of the breech ring. It protrudes into the breechblock cut of the breech ring to contact the upper end of the trigger when the gun is fired.

9. Disassembly and Assembly of the Gun, 76-mm, M32

a. General. The disassembly and assembly procedures outlined herein are recommended for guidance of the crew when performing normal care and cleaning, and for inspection and replacement of parts. Two crew members should work together in removing the breechblock.

b. Field Disassembly for Daily Maintenance or After Firing. The only tool necessary is a heavy-duty screwdriver.

(1) *Inspection.* Insure that the breechblock crank stop is in its normal rear position. Open the breech far enough

for inspection of the chamber, and make certain the gun is not loaded. Close the breech, and actuate the firing mechanism to release the compression of the firing spring.

(2) *Firing spring retainer.* Push the firing spring retainer in, and rotate it until the slot in the retainer is vertical; then release it, allowing the retainer to be pushed out into the hand. Withdraw the firing spring.

(3) *Percussion mechanism.* Cup one hand over the percussion mechanism well, and pull the cocking lever to the rear. The percussion mechanism will fall into the cupped hand.

Note. Disassembly of the percussion mechanism should be performed *only* for inspection or when a malfunction or defective part makes it necessary. Procedure is as follows: Press the firing spring stop into the firing pin guide, and drift out the firing pin retaining pin. Unscrew the firing pin from the guide, releasing the retracting spring and the firing spring stop (fig. 10).

(4) *Breechblock crank stop.* Depress the gun tube slightly; then, using an improvised drift, push up on the breechblock crank stop plunger and slide the stop forward.

(5) *Breechblock.* One crew member will unlatch the operating handle and rotate it to the rear. At the same time, a second crew member will grasp the lower part of the breechblock as it moves downward. When the crossheads move out of the inclined T-slot of the breechblock, remove the breechblock from the breech ring. Return the operating handle to the latched position.

(6) *Crossheads and extractors.* Remove the crossheads from the pivot on the breechblock crank. Remove the right and left extractors from the breech ring.

(7) *Extractor plunger plugs, plunger springs, plungers, and links.* Using a heavy-duty screwdriver, unscrew the extractor plunger plugs and remove the plugs, extractor plunger springs, and extractor plungers. Remove the extractor links.

(8) *Cocking parts.* With the breechblock on its front face, remove the cocking lever from the left side of the breechblock. Using a screwdriver, depress the cocking bar pivot plunger and rotate it until the stud enters the slot opening in the rear face of the breechblock. Remove the cocking bar pivot and cocking bar.

(9) *Firing parts.* Remove the trigger, sear, and sear spring from the right side of the breechblock.

c. *Field Assembly for Daily Maintenance or After Firing.*

(1) *Firing parts.* Replace the sear spring and sear into the

sear recess, then replace the trigger. (Two holes are provided in the sear for adjustment of tension of the sear spring.) (Difficulty in replacing the sear can be avoided by inserting a tang of the sear spring into its recess in the breechblock until it is barely caught. Insert the sear into its recess, checking to be sure that the other tang on the sear spring enters a recess in the sear.)

(2) *Cocking parts.* Replace the cocking bar in its recess with the flat projection first and the slotted side toward the top of the breechblock; replace the cocking bar pivot, and check to make sure that the two are engaged. Rotate the cocking bar pivot plunger until the stud is disengaged. Replace the cocking lever, and check to see that all parts operate properly.

(3) *Extractor links, plungers, plunger springs, and plunger plugs.* Replace the extractor links in their recesses in the breech ring so that the cutaway part is to the rear and toward the breechblock recess. Replace the extractor plungers, plunger springs, and plunger plugs. Screw the extractor plunger plugs in securely with a screwdriver. (Extractor links are interchangeable.)

(4) *Crossheads and extractors.* Replace the extractors in the breech ring. Replace the crossheads on the pivot of the breechblock crank. (Crossheads are interchangeable; extractors are not interchangeable.)

(5) *Breechblock.* One crew member will unlatch and pull to the rear and down on the operating handle, rotating the breechblock crank as far as possible. A second crew member will hold the upper arm of the cocking lever to the rear and insert the breechblock into the lower portion of the breech ring. Raise the breechblock until the top of the trigger is even with the loading notch in the breech ring. Allow the breechblock crank to rotate so that the crossheads move into the inclined T-slot. Return the operating handle to the latched position. Trip the extractors with the extracting and ramming tool to permit the breechblock to move to the closed position.

Caution: Never trip the extractors with the fingers.

(6) *Breechblock crank stop.* Slide the breechblock crank stop to the rear. Check to make certain that the breechblock crank stop is locked in the rear position by the plunger.

(7) *Percussion mechanism.* Replace the percussion mechanism in the percussion mechanism well. Move the trigger to the rear to allow the collar of the percussion mechanism to pass the sear. Replace the firing spring.

(8) *Firing spring retainer.* Place the cupped end of the firing spring retainer over the firing spring. With the slot vertical, push in and rotate the retainer until the slot is horizontal.

(9) *Test.* Cock and actuate the firing mechanism.

d. *Disassembly During Weekly Maintenance.* Tools necessary are a heavy-duty screwdriver, an 8-inch improvised drift ($3/8$-in. or less in diameter), a closing spring housing wrench, and a 12-inch crescent wrench.

(1) *Field disassembly.* Complete the field disassembly for daily maintenance (*b*, above).

(2) *Cocking bar pivot plunger and spring.* Rotate the plunger until the stud clears the notch, and remove the cocking bar pivot plunger and spring.

(3) *Closing spring.* By use of a screwdriver, loosen the closing spring housing retainer screw. Using the closing spring housing wrench, rotate the housing slightly to release the tension against the housing plunger. Using a drift, push the plunger in beyond the housing notches, and allow the housing to rotate, relieving the tension of the closing spring. Rotate the operating shaft crank until the lug is to the rear.

(4) *Left recoil guard.* Using a 12-inch crescent wrench, remove the two self-locking nuts from the left recoil guard. Remove the guard. (Remove the lower lock nut first.)

(5) *Operating shaft.* Using a drift, push in on the operating shaft plunger and remove the operating shaft retainer. As the operating shaft is moved to the left, remove (in order) the spacer, breechblock crank, and closing spring. Completely remove the operating shaft. Further loosen the closing spring retainer screw and remove the closing spring housing.

e. *Assembly During Weekly Maintenance.*

(1) *Operating shaft.* Replace the closing spring housing; tighten the closing spring retainer screw slightly. Slide the operating shaft through the closing spring housing; and, as the shaft is moved to the right, replace the closing spring, breechblock crank, and spacer. After completion of the above, move the operating shaft farther to the right and engage the operating shaft in the operating crank. Replace the operating shaft retainer.

(2) *Left recoil guard.* Replace and tighten the self-locking nuts securely.

(3) *Closing spring.* Rotate the breechblock crank up into the breech ring as far as possible. Tighten the closing spring housing retaining screw. Push in on the closing spring housing plunger and, by use of a closing spring housing wrench, rotate the housing until the plunger engages the third notch.

(4) *Cocking bar pivot spring and plunger.* Push the plunger into its recess, and rotate it until the stud enters the slot opening of the breechblock.

(5) *Complete field assembly.* See paragraph 9c.

Note. Disassembly other than that explained in paragraphs 9d and 9e will be accomplished only under the supervision of the turret mechanic.

10. Functioning of the Gun, 76-mm, M32

a. Manual Opening of the Breech. Grasp, unlatch, and pull to the rear on the operating handle. As the operating handle is rotated to the rear, the lug on the operating handle hub contacts a similar lug on the operating shaft crank (fig. 8). The operating shaft crank, in turn, rotates to the rear and, being locked by a cap screw to the operating shaft, rotates the operating shaft, breechblock crank, and operating crank. The rotation of the breechblock crank winds the closing spring, increasing its tension. The crossheads of the breechblock crank, riding in the inclined T-slot of the breechblock, move the breechblock to the open position. As the breech opens, the trunnions of the extractors, moving in the curved extractor grooves of the breechblock, are forced forward; and the flat surfaces of the trunnions are positioned directly above the trunnion seats of the breechblock. When the operating handle is rotated forward, the closing spring unwinds slightly, moving the breechblock upward until the trunnion seats contact the flat surfaces of the trunnion and lock the breechblock in the open position. The extractor plunger springs expand to insure locking of the breechblock in the open position. Return the operating handle to the latched position.

Caution: The operating handle should be raised and latched immediately after manual opening of the breech and should never rest in an unlatched position.

b. Cocking. Cocking occurs as the breechblock is moved to the open position (fig. 6). As the breechblock moves to the open position, the upper arm of the cocking lever is cammed rearward by a camming surface of the breech ring. The lower arm of the cocking lever moves forward against the outer arm of the cocking bar. The cocking bar rotates the cocking bar pivot, depressing the pivot plunger and compressing the pivot plunger spring. The inner arm of the cocking bar moves to the rear, contacting the collar of the percussion mechanism. This moves the percussion mechanism to the rear, compressing the firing spring between the firing spring stop and the firing spring retainer. The rearward movement of the percussion mechanism rotates the sear and winds the sear spring by engaging the flat portion of the sear. When the percussion mechanism is far enough to the rear for the collar to clear the sear, the sear spring rotates the sear to its original position, blocking the percussion mechanism, thus preventing the firing spring from expanding.

c. Loading (Automatic Closing of Breech). The breechblock moves automatically to the closed position when a round is loaded into the chamber. The rim of the cartridge case contacts the lips of the extractors, forcing them forward and moving the inner trunnions of the extractors rearward off the extractor trunnion seats of the breechblock and into the curved extractor grooves. As this action occurs, the extractor plunger springs are compressed. The closing spring then unwinds, rotating the breechblock crank, operating shaft, operating crank, and operating shaft crank, moving the breechblock to the closed position, thus chambering the round. The cocking bar pivot plunger, moved by expansion of the pivot plunger spring, rotates the pivot, cocking bar, and cocking lever to the normal position. This moves the cocking bar clear of the path of the percussion mechanism.

d. Firing. Firing of the 76-mm gun may be accomplished electrically or manually. The same parts of the breech mechanism are involved in both methods. (See paragraph 19 for explanation of electrical and manual firing procedure.) When the trigger plunger is moved rearward, it contacts the upper arm of the trigger and pushes it rearward (figs. 8 and 9). The lower arm of the trigger moves forward, engages an offset lug on the sear, and rotates the sear. When the sear is rotated, the flat portion of the sear clears the collar on the percussion mechanism and allows the percussion mechanism to move forward under the action of the firing spring. When the firing spring stop strikes the inner face of the breechblock, the expansion of the firing spring is halted. The firing pin and guide continue forward under inertia. This action compresses the retracting spring between the firing spring stop and the head of the firing pin. The forward motion of the guide and the pin ceases when the firing pin strikes the primer. The retracting spring then expands, withdrawing the firing pin from the primer and into the firing pin well.

e. Automatic Opening of the Breech. When the gun recoils, the lug on the operating crank moves the operating cam outward, compressing the operating cam return spring. As the lug clears the cam, the return spring expands, moving the cam to its original position. As the gun counterrecoils, the lug on the operating crank strikes the operating cam and is rotated to the rear. As the operating crank rotates, it rotates the operating shaft, breechblock crank, and operating shaft crank, moving the breechblock to the open position. At the same time, the closing spring is being wound. As the breech opens, the trunnions of the extractors, moving in the curved extractor grooves of the breechblock, are forced forward, and the flat surfaces of the trunnions are positioned directly above the trunnion seats of the breechblock. The closing spring unwinds slightly, moving the breechblock upward until the trunnion seats contact the flat surfaces of the trunnions. This locks the breechblock in the open position. The extractor plunger springs expand to insure locking of the breechblock in the open position.

f. Extraction and Ejection. Extraction and ejection occur when the breechblock is automatically opened during counterrecoil. As the breechblock nears the fully open position and has cleared the rear of the cartridge case, the extractor trunnions are cammed forward by the curved extractor grooves, rotating the extractor lips to the rear. The lips of the extractors, being in front of the rim of the cartridge case, extract the case from the chamber and eject it from the breech ring.

11. Recoil Mechanism, Gun, 76-mm, M32

a. General. Major components of the concentric, hydrospring type, recoil mechanism are the recoil cylinder assembly and the replenisher assembly.

b. The Recoil Cylinder Assembly.
  (1) The recoil cylinder assembly (fig. 13) is composed of the cradle, gun tube, recoil piston, and counterrecoil spring. The recoil cylinder is formed by the inside of the cradle and the outside of the gun tube inclosed by the cradle. The recoil piston is keyed to the gun tube near the front of the cradle. The counterrecoil spring is coiled around the gun tube between the recoil piston and the rear of the cradle. When ready for operation, the recoil cylinder is full of hydraulic oil.
  (2) When the gun is fired, it recoils. As it moves to the rear in recoil, the hydraulic fluid is forced from the rear of the piston to the front of the piston; and at the same time, the counterrecoil spring is being compressed. The inside of the cradle is tapered inward from front to rear so that the clearance between the piston and cradle is greatest at the beginning of recoil. As the gun nears the end

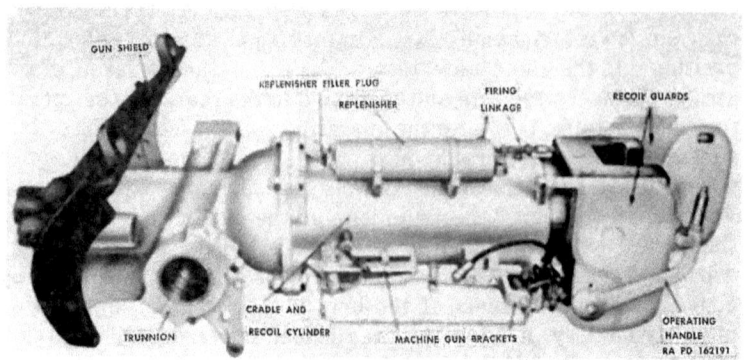

Figure 13. Gun mount (left side).

of recoil, the throttling of the flow of the hydraulic oil by the taper absorbs the force, halting rearward movement. Some of the force of recoil is used to compress the counterrecoil spring. When recoil is completed, the compressed counterrecoil spring expands, moving the gun forward. This movement forces the oil from the front of the recoil piston to the rear of the piston. Near the end of counterrecoil, an enlarged portion of the gun tube enters the buffer chamber in the front of the recoil cylinder. The movement of oil out of the buffer chamber is thereby restricted and the cushioning effect of the restricted flow eases the gun into battery without undue shock.

Note. A buffer chamber is provided in the recoil mechanism which permits correct throttling action and return of the gun to battery position, even though the viscosity of the hydraulic oil changes. It consists of a bypass port from the buffer chamber to the recoil cylinder and a buffer regulator that adjusts the size of the port through which the hydraulic oil passes. If the gun is too slow in returning to battery, or if it slams into battery, the size of the bypass should be increased or decreased as necessary. This adjustment is authorized to be made only by ordanance personnel.

c. *The Replenisher Assembly.*
  (1) The replenisher assembly (fig. 14) consists of the replenisher cylinder, replenisher piston, replenisher piston spring, indicator assembly, a spring-loaded ball valve, and a hose connection. The replenisher is connected to the recoil cylinder by a flexible hose.
  (2) The recoil oil under spring pressure in the replenisher compensates for expansion and contraction of oil in the recoil mechanism due to changes in temperature. The

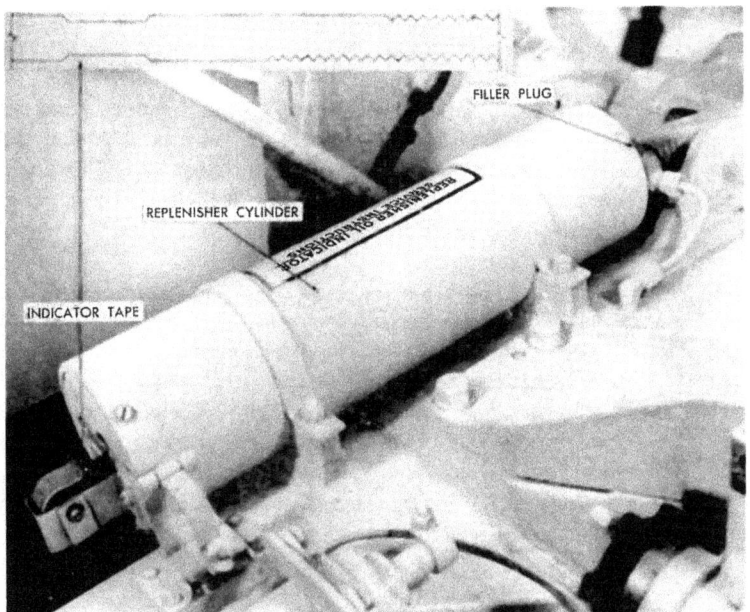

Figure 14. Replenisher.

expanding oil moves through the flexible hose from the recoil cylinder, the replenisher compressing the replenisher spring and causing the indicator tape to wind around the screw in the indicator assembly. The steel indicator tape is designed to be read by touch. Both edges rough indicates that the system is dangerously low on oil and should be refilled; one edge rough and one edge smooth indicates normal operating range with the proper amount of oil in the system; both edges smooth indicates an excess of oil, which must be removed.

d. *Checking, Filling, and Draining.*
  (1) The recoil system is checked by feeling the indicator tape on the replenisher (par. 10c(2)). This check is performed before firing while the recoil system is cool.
    Caution: Pull on the tape to insure that it is not broken.
  (2) If the amount of oil in the system is not correct, proceed as follows:
    (a) If there is not enough oil in the system, remove the filler plug from the replenisher and the nozzle from the filler gun. Fill the gun with the proper oil, screw the filler gun hose into the filler plug hole loosely, push on the

plunger to force the air out, screw the hose tight, and force the oil into the replenisher. Repeat this process until the indicator tape shows the proper amount of oil.

(b) If there is too much oil in the system, remove the filler plug, hold a rag under the filler plug hole, and drain the excess oil from the replenisher onto a rag. To drain the excess oil, it is necessary to push in on the spring-loaded ball valve in the filler plug hole with the nozzle of the filler gun.

Note. Use gradual pressure on the ball valve to control the amount of oil flowing from the replenisher. Check the indicator tape frequently to prevent draining too much oil.

(3) During firing, the system should be checked periodically for leaks or excessive oil due to overheating. After several rounds have been fired, the indicator tape may show both edges smooth as a result of the recoil oil expanding. It is not necessary to drain any oil from the replenisher when this occurs. However, if the long notches are exposed, the crew should drain enough oil so that the two smooth edges are exposed.

e. Bleeding. If the recoil of the gun is excessive and the replenisher shows the correct amount of oil, there may be air in the recoil system. To correct this, the turret mechanic should bleed the recoil system.

12. Malfunctions, Gun, 76-mm, M32

a. General. A malfunction is an unintentional cessation of fire caused by a failure of some part of the gun, the mount, or the ammunition. Malfunctions are divided into four general classes: failure to load, premature firing, failure to fire, and failure to extract and eject. The malfunctions discussed here are the most common ones but do not necessarily include all malfunctions which may occur.

b. Failure To Load (Breech Will Not Fully Close).

Cause	Correction
Dirty, bulged, or dented round.	Remove round and load another.
Dirty chamber.	Remove round and clean chamber.
Improper adjustment of closing spring.	Remove round and increase the tension on the closing spring (par. 9e(3)).

c. Premature Firing (Gun Fires as Breech Closes).

Cause	Correction
Weak sear spring causing sear to release percussion mechanism as breechblock closes.	Replace defective parts.

d. Failure To Fire. (Check to Determine Whether the Manual Safety is Off.)

Cause	Correction
Gun out of battery (too much oil in recoil system).	Remove excess oil from replenisher.
Defective primer.	Remove round, load another round, and attempt to fire.
Worn or broken firing pin, firing spring, sear, or sear spring.	Replace defective part.
Trigger is broken.	Replace defective part.
Dirt, excess grease, or wax in percussion mechanism well, preventing free movement of precussion mechanism.	Clean percussion mechanism and well.
Worn or broken cocking lever, cocking bar, cocking bar pivot, or pivot plunger spring.	Replace defective part.

e. Failure To Extract and Eject.

Cause	Correction
Defective rim of cartridge case causes extractor lips to pull through rim, leaving case in chamber.	Pry or ram case out of chamber.
Broken, weak, or missing extractor plunger springs permit extractor trunnions to slip off trunnion seats of breechblock, allowing breechblock to jam cartridge case in breech ring.	Replace missing or defective part.

13. Care, Cleaning, and Lubrication, Gun, 76-mm, M32

　a. General.

　　(1) It is vitally important to keep all materiel in proper condition for immediate service. Tools, accessories, and lubricating, cleaning, and preserving materials are provided for this purpose.

　　(2) Proper lubrication with the proper lubricants and at specified intervals is essential to the care and preservation of the materiel. Refer to LO 9–7016 for a lubrication guide.

　　(3) All protective covers for the gun and mount should be installed when the gun is not in service. Exposed unpainted surfaces are to be cleaned, dried, and covered with a coat of rust preventive compound medium. If the materiel is not to be used for an extended period, process it in accordance with current supply bulletins.

　　(4) During disassembly and assembly, thoroughly clean disassembled parts before oiling and assembling them. Do

not use a steel hammer directly on any part. If a copper or lead hammer is not available, use a wood block as a buffer. Always use the tool intended for the part. Tools and accessories, as well as the materiel, should be kept clean, free from rust, and protected with preservative oil as required. Crocus cloth is the only abrasive authorized for removing rust.

(5) The quadrant seats on the breech ring must be protected. Do not place tools or other articles upon them.

b. *Before Firing.*
   (1) *Tube.* Check bore and chamber for dirt, defects, or obstructions. Clean and wipe dry.
   (2) *Breech mechanism.* Examine the breech mechanism for functioning and cleanliness. Clean and lubricate as necessary.

c. *During Firing.* Be alert for any malfunction. Lubricate as necessary.

d. *After Firing.*
   (1) *Tube.* After firing and for three consecutive days thereafter, thoroughly clean the bore with rifle bore cleaner, making sure that all surfaces in the bore and chamber are well coated. Wipe the bore dry each time before applying additional bore cleaner. After the fourth cleaning, dry the bore and oil it with the prescribed lubricant unless the gun will be fired in the next 24 hours. When the gun is not being fired, renew the oil film weekly.
   (2) *Breech mechanism.* Disassemble, clean with bore cleaner, lubricate, and assemble. Check for proper functioning and condition of parts.
   (3) *Bore evacuator* (fig. 5). The bore evacuator should be cleaned each firing day and at least once each week when not fired. More frequent cleaning may be necessary if the residual gases from the firing are not cleared from the tube by the bore evacuator. To clean the bore evacuator, remove the blast deflector and evacuator can. Clean the jets in the gun tube by inserting a piece of wire through each of them. Do not remove the jet inserts (if so equipped) from the gun tube during crew maintenance. Clean the unpainted exterior of the tube and the interior of the evacuator can with bore cleaner. Coat the cleaned parts with the prescribed preservative, and reinstall them.

   Caution: Be careful not to damage the sealing lips while removing and replacing the bore evacuator.

Section III. MACHINEGUN MOUNTS

14. General

a. The firepower of the caliber .50 and caliber .30 machineguns greatly increases the shock effect of the M41 and M41A1 tanks.

b. This section describes and illustrates the mounts for these guns and furnishes essential information for the crew to mount and operate them.

15. Combination Gun Mounts, M76 and M76A1

a. General. Each of these combination gun mounts will mount either an M1919A4E1 or M37 Machinegun, Caliber .30, coaxially with the 76-mm gun (fig. 13). The machinegun mount is secured to the tank gun cradle and is part of the tank gun mount. When the tank gun is elevated and traversed, the machinegun is moved correspondingly.

b. Location of Machinegun Mount. The coaxial machinegun mount is located on the left side of the combination gun mount. Two bracket assemblies, front and rear, hold the machinegun when mounted (fig. 13).

c. Components. The component parts of the combination gun mount are—
  (1) Front mounting bracket.
  (2) Machinegun front locking pin.
  (3) Coaxial machinegun cradle.
  (4) Elevating and traversing mechanism.
  (5) Machinegun rear locking pin.

d. Installation of Coaxial Machinegun.
  (1) Pull out the machinegun front and rear locking pins.
  (2) Insert the muzzle of the machinegun into the port of the gun shield.
  (3) Set the machinegun into the front mounting bracket, aline the front mounting holes, and insert the front locking pin.
  (4) Pull up the elevating and traversing mechanism, and aline it with the rear mounting holes of the machinegun.
  (5) Insert the rear locking pin, and the machinegun will be secured in the coaxial gun mount.
  Note. Check bolts on front mounting bracket frequently for tightness.

e. Adjustment of the Firing Solenoid (fig. 16). (Headspace must be adjusted prior to adjustment of the firing solenoid.)
  (1) The adjusting screw is located at the bottom of the solenoid support, between the solenoid and the machinegun

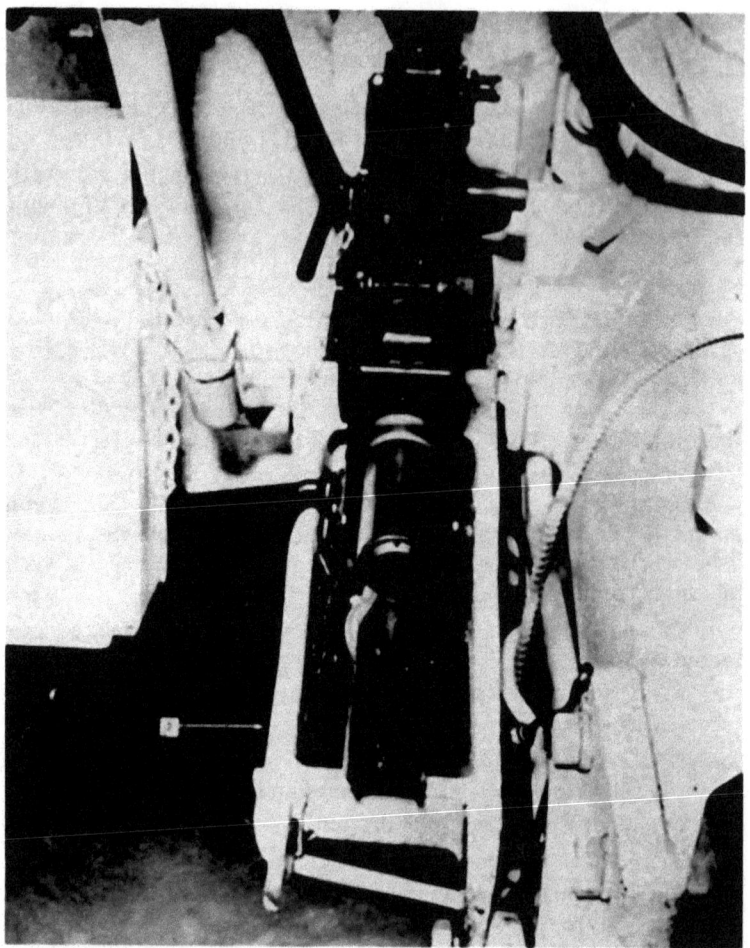

Figure 15. Coaxial machinegun, mounted.

elevating bracket. The lock screw is located on the machinegun elevating bracket.

(2) Loosen the jam nuts with a $7/16$-inch open-end wrench. Loosen the lock screw with a screwdriver.

(3) Cock the gun and, using a screwdriver, turn the adjusting screw clockwise to move the solenoid plunger away from the trigger of the machinegun; turn the adjusting screw counterclockwise to move the solenoid plunger closer to the trigger of the machinegun.

(4) Adjustment is correct when there is a $1/32$-inch clearance

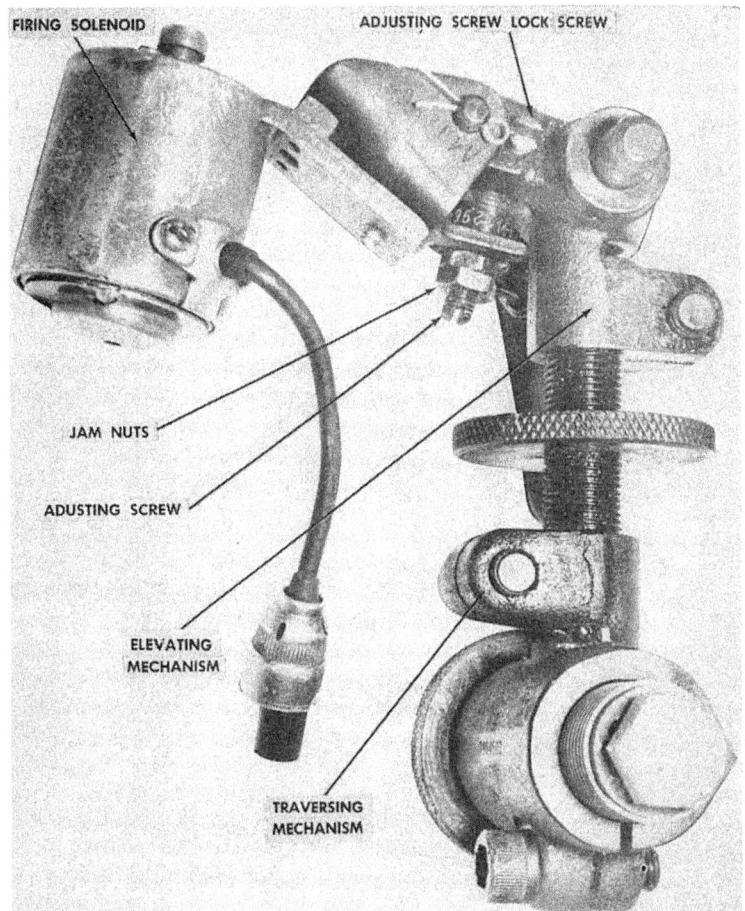

Figure 16. Machinegun elevating and traversing mechanism, showing solenoid.

between the solenoid plunger and the trigger of the machinegun.

(5) When the adjustment is completed, tighten the lock screw and the jam nuts on the solenoid adjusting screw. Check for final adjustment by holding the electric firing trigger while manually cocking the weapon several times to insure that the firing pin releases as the bolt moves into battery.

f. *Loading and Firing the Coaxial Machinegun.*

(1) Load the magazine, and run the belt through the feedway and into the machinegun.

(2) To half-load, pull the retracting slide to the rear and release it. To full-load, repeat the operation.

(3) To fire electrically:

   (*a*) Turn on the caliber .30 machinegun switch.

   (*b*) Press either of the gunner's firing triggers or the commander's firing trigger.

g. Boresighting the Coaxial Machinegun. The coaxial machinegun must be alined with the tank gun. The back plate and bolt groups must be removed to boresight.

(1) To adjust the machinegun for elevation:

   (*a*) Loosen the two upper socket-head cap screws.

   (*b*) Turn the elevating adjusting screw clockwise to elevate the muzzle of the machinegun; turn the screw counterclockwise to depress the muzzle of the machinegun.

(2) To adjust the machinegun for traverse:

   (*a*) Loosen the lower socket-head cap screw.

   (*b*) Turn the top of the traverse adjusting screw toward you to move the machinegun muzzle to the right; turn the top of the traverse adjusting screw away from you to move the machinegun muzzle to the left.

(3) Sight through the bore of the gun, and aline the axis of the bore on the boresight point by means of the elevating and traverse adjusting screws.

(4) Tighten the socket-head cap screws, and reassemble the gun.

h. Removal of Coaxial Machinegun.

(1) Clear the machinegun.

(2) Remove the front and rear locking pins, and lift out the machinegun.

16. Caliber .50 Machinegun, M2, HB, Turret-Mounted

a. General. The Machinegun, Caliber .50, M2, HB, is located on the turret roof adjacent to the loader's hatch. The turret-mounted gun is used against both ground and air targets and is controlled and fired manually. The machinegun and the cradle assembly may be removed as a unit. When the machinegun is removed, replace the cover plug assembly. When the tank is in motion, the machinegun may be locked in place by the machinegun traveling lock located forward of the commander's cupola.

b. Components. The pintle mount consists of—

(1) Pintle stand assembly.

(2) Lock handle.

(3) Cover plug assembly.
(4) Cradle assembly.
(5) Front and rear locking pins.

c. *Installation.*
(1) Remove the cover plug assembly.
(2) Place the pintle lock in the open position (handle down).
(3) Insert the cradle assembly in the pintle stand and lock it (handle up).
(4) Aline the front and rear machinegun mounting holes with holes on the cradle assembly.
(5) Insert the front and rear locking pins.

d. *Removal.*
(1) Clear the machinegun.
(2) Reverse the procedure for installation.

17. Tripod Mount, M3

Tripod Mount, M3, is provided for the caliber .50 machinegun. The tripod mount is stored in the left front fender box when not in use.

18. Maintenance of Machinegun Mounts

Maintenance of the combination gun mount and pintle mount, other than inspection and normal care, cleaning, and lubrication, by the using arm is not authorized.

Section IV. TURRET AND ARMAMENT CONTROLS AND EQUIPMENT, M41 TANK

19. General

a. The principal differences in the Tanks, 76-mm Gun, M41 and M41A1, are found in the turret and armament controls and equipment. This section covers the controls and equipment found on the M41; section V covers the controls and equipment of the M41A1. For steps in placing the turret into power operation, for both tanks, see paragraph 34.

b. The turret and armament controls and equipment are those devices used by the tank commander, gunner, and loader for accurate and rapid employment of the tank's armament. Traversing controls permit laying the gun for direction through rotation of the turret on the hull. Elevating controls permit laying the gun for elevation by pivoting the gun and mount on trunnions located in the forward turret wall. The gunner has both power and manual control of traversing, whereas the tank commander has only power traverse. The loader has a traverse safety which, when off, will prevent power traverse of the turret while the loader is removing ammunition from the stowage rack to the right of the driver's compartment. Firing controls provide the gunner with gun switches, electrical firing triggers, a hand firing lever for the main armament, and safeties to prevent accidental discharge. The tank commander's firing controls operate electrically only. Auxiliary turret equipment provides the turret crew members with certain safety devices, stowage brackets, and turret and gun locks which contribute to the operating efficiency of the tank.

20. Gunner's Power Control, M41 Tank

The gunner's power control handle (fig. 17) is located directly in front of the gunner's position. It provides the gunner with power control of the turret. The turret can be traversed 360° in either direction as follows: to traverse right, rotate the control handle clockwise; to traverse left, rotate the control handle counterclockwise. The amount the handle is rotated from neutral determines the speed of turret traverse.

21. Tank Commander's Power Control, M41 Tank

a. The tank commander's power control handle (fig. 18) is located on the turret wall to the tank commander's right front and gives him the same power control of the turret that the gunner has. In addition, the commander's control handle is equipped with an override lever.

b. To traverse the turret, press the override lever, and rotate the power control handle in the desired direction of traverse. The amount the handle is moved from neutral determines the speed of traverse. By pressing the override lever on his control handle, the tank commander takes the power control of the turret from the gunner.

22. Loader's Traverse Safety, M41 Tank

a. The loader does not have controls which will permit him to traverse the turret. For his safety, however, while he is stowing ammunition or removing ammunition from the stowage compartment, a loader's traverse safety switch and indicator light (fig. 19) are installed on the turret roof behind his position.

b. When the traverse safety switch is turned to the ON position, the indicator light should glow, and the gunner and tank commander can traverse the turret in power. When the switch is in the OFF position, the turret will not traverse in power.

Caution: The loader should never attempt to remove ammunition from the ammunition rack to the right of the driver's compartment when the indicator light is on.

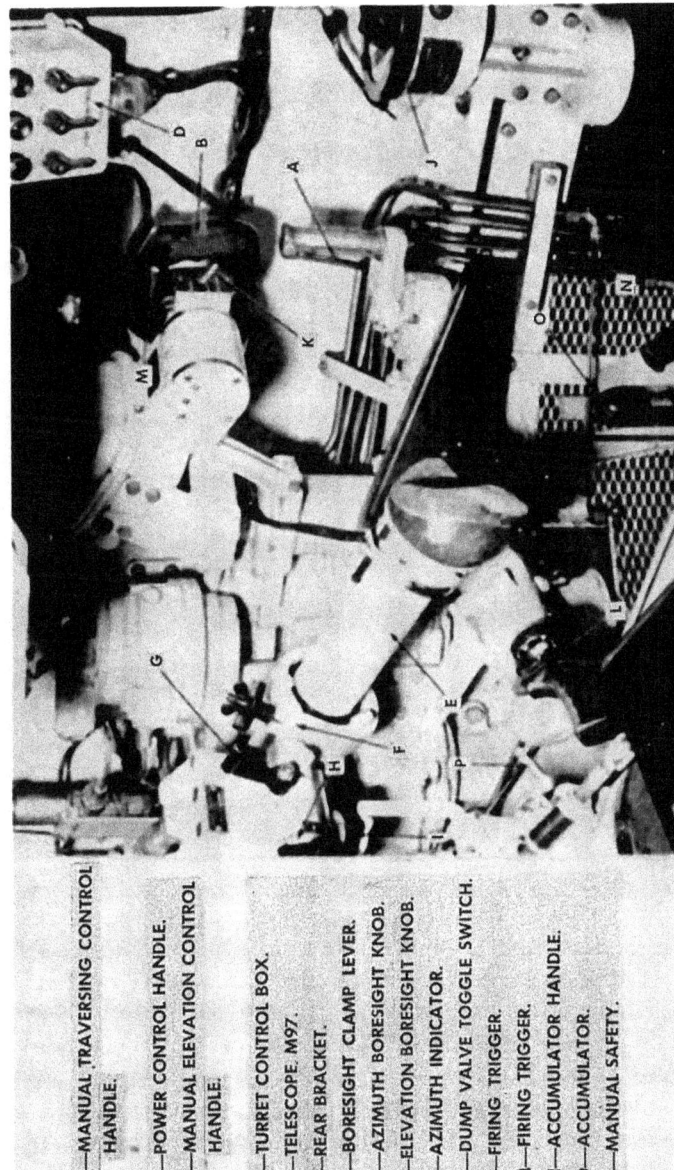

A—MANUAL TRAVERSING CONTROL HANDLE.
B—POWER CONTROL HANDLE.
C—MANUAL ELEVATION CONTROL HANDLE.
D—TURRET CONTROL BOX.
E—TELESCOPE, M97.
F—REAR BRACKET.
G—BORESIGHT CLAMP LEVER.
H—AZIMUTH BORESIGHT KNOB.
I—ELEVATION BORESIGHT KNOB.
J—AZIMUTH INDICATOR.
K—DUMP VALVE TOGGLE SWITCH.
L—FIRING TRIGGER.
M—FIRING TRIGGER.
N—ACCUMULATOR HANDLE.
O—ACCUMULATOR.
P—MANUAL SAFETY.

Figure 17. Gunner's controls, M41.

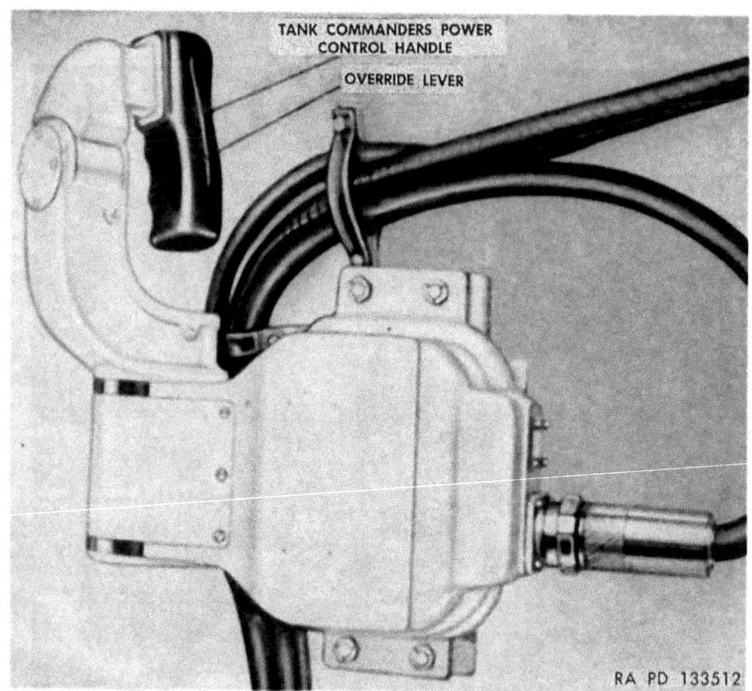

Figure 18. Tank commander's power control handle, M41.

23. Gunner's Manual Controls, M41, Tank

a. Manual Traverse Control Handle. The gunner's manual traverse control handle (fig. 17) is located above and to the right of the gunner's seat. To traverse the turret manually, grasp the manual traverse control handle, squeeze the release lever on the handle, and rotate the handle in the desired direction of traverse. The speed of traverse is regulated by the rate of handle rotation. Four recesses spaced at 90° intervals provide four positions for locking the manual traverse control handle. A no-back mechanism automatically holds the turret in position and prevents turret drift.

b. Dump Valve Toggle Switch (fig. 17). A toggle switch is provided on the gunner's power control assembly to enable the gunner to traverse manually while the turret motor is running. It is located to the left of the gunner's power control handle. With the turret motor running and with the switch in the POWER position, the turret can be traversed in power only. When the switch is in the MANUAL position, the turret can be traversed manually only.

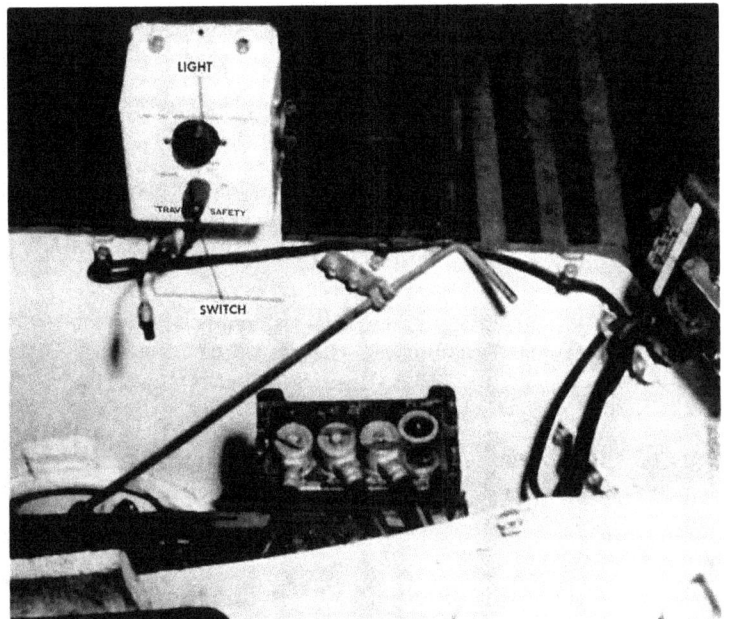

Figure 19. Loader's traverse safety switch, M41.

c. Manual Elevation Control Handle. The gunner's manual elevation control handle (fig. 17) is located to the left front of the gunner's position. It is connected to a hydraulic pump which, by directing a flow of oil into an elevation cylinder either below or above a piston, causes the gun to be elevated or depressed. To elevate the gun, rotate the handle clockwise. To depress the gun, rotate the handle counterclockwise.

d. Accumulator Hand Pump and Handle (fig. 17). A hand pump is located to the right of the gunner. This pump is used to charge the manual elevation system and is operated by means of a handle located to the right of the gunner. To charge the system, move the handle up and down until the gun responds quickly to movements of the manual elevation control handle.

24. Turret Control Box, M41 Tank

a. The turret control box (fig. 17) is located on the turret wall to the right of the gunner's position. On it are mounted switches and indicator lights that control the turret power and the electrical firing circuits.

b. The turret motor switch (AUTOMATIC) is on the right side of the turret control box. This switch must be in the ON position before the turret can be operated in power. When the switch is on, the indicator light above it should be on.

Note. When the turret motor is running, the noise level is quite high because of characteristics inherent in the system. This is not a cause for alarm if the system has been properly serviced and maintained.

c. The 76-mm gun switch is in the center of the turret control box. This switch controls the electrical firing circuit for the main armament and must be in the ON position before the 76-mm gun can be fired electrically. When the 76-mm gun switch is on, the indicator light above it should be on.

d. The coaxial machinegun switch is on the left side of the turret control box. This switch controls the electrical firing circuit for the coaxial machinegun. When the coaxial machinegun switch is on, the indicator light above it should be on.

25. Firing Controls, M41 Tank

a. Electrical. The master relay switch and the appropriate gun switch must be on in order to fire either the 76-mm gun or the coaxial machinegun electrically.

  (1) The gunner may fire the 76-mm gun or the coaxial machinegun electrically by using either one of the firing triggers (fig. 17), which are located as follows: one on the handle grip of the manual elevation control handle, and one on the handle grip of the gunner's power control handle.

  (2) The tank commander may fire the 76-mm gun or the coaxial machinegun electrically by using the firing trigger on the front of the tank commander's power control handle. This trigger will function only when the turret motor and the proper gun switch are on and the tank commander has taken power control of the turret by pressing the override lever of his power control handle.

  (3) The loader's reset safety (fig. 20) is located on the left front portion of the turret wall and is connected to a microswitch near the breech ring on the left side of the gun mount. The microswitch operates to open and close the electrical firing circuit of the 76-mm gun. The circuit is broken when the gun recoils, turning off the indicator light on the reset safety box. The safety is reset and the circuit completed again by pushing the button. The loader must perform this operation after each round is fired. When the circuit is closed and ready for firing, the indicator light on the box should be on and the gun-ready lights at all direct-fire sights should glow.

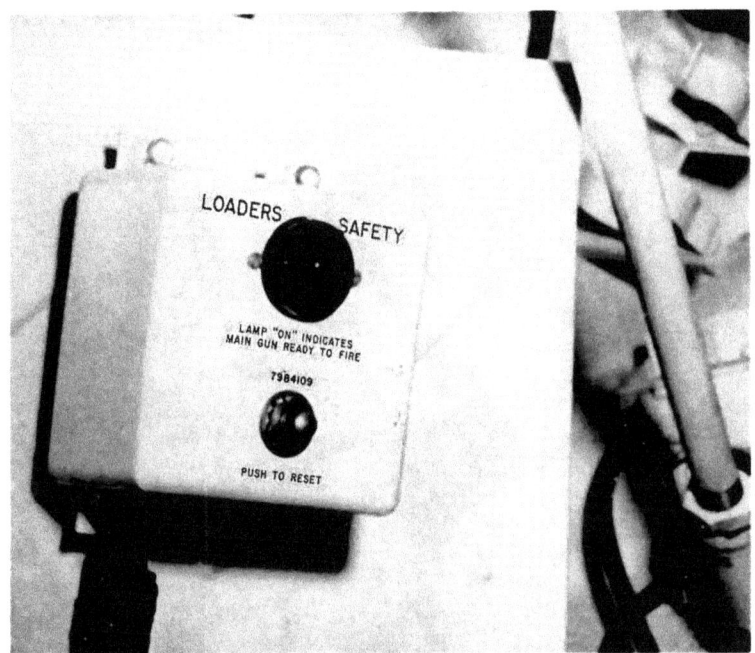

Figure 20. Loader's reset safety.

b. Manual. When the gun is being fired manually, no electrical firing circuits are used. Manual firing of the 76-mm gun is accomplished by rotating the hand firing lever (fig. 21) forward. The cam on the hand firing lever rotates against the end plate of the firing plunger, forcing it against the trigger plunger.

Caution: Release the hand firing lever immediately after the gun fires.

c. Safeties.
  (1) *Timing relay (electrical).* When the 76-mm gun is being fired electrically, the timing relay breaks the electrical circuit immediately after the firing plunger is actuated by the solenoid. As the gun recoils, the circuit is broken by the microswitch ($a(3)$, above). Before the 76-mm gun can be fired again it is necessary for the gunner (or tank commander) to release the firing trigger and actuate it again, *after* the loader has reset the safety.
  (2) *Manual safety.* The 76-mm gun manual safety (fig. 21) is located on the firing lever bracket. When the safety is behind the end plate of the firing plunger, it is on SAFE and prevents electrical or manual firing of the 76-mm gun. When the safety is down, the gun can be fired.

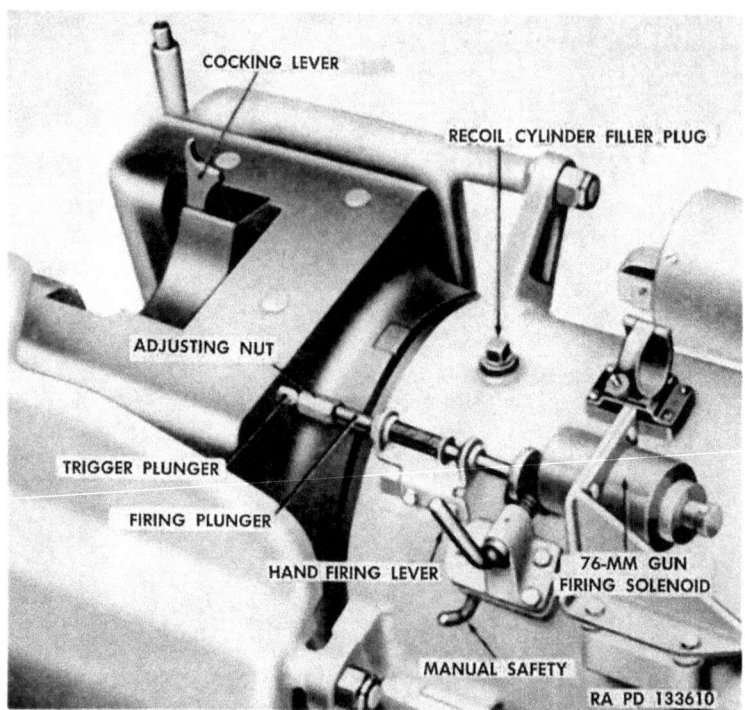

Figure 21. Firing linkage.

26. Turret and Gun Traveling Locks, M41 Tank

a. Turret Lock.

(1) The turret lock (fig. 22), located to the right rear of the gunner's seat, holds the turret stationary by means of a gear segment which engages the turret ring gear teeth. The turret lock should be in the locked position while the vehicle is in motion, unless the turret is to be traversed or the gun is in the gun traveling lock.

(2) To unlock the turret, turn the turret lock handle clockwise until the gear segment disengages from the turret ring gear and the turret can be traversed.

(3) To lock the turret, turn the handle counterclockwise until the turret cannot be traversed.

Caution: Use manual traverse to check the turret lock.

b. Gun Traveling Lock.

(1) The gun traveling lock, located on the left rear fender of the tank, is used to keep the gun in the locked position

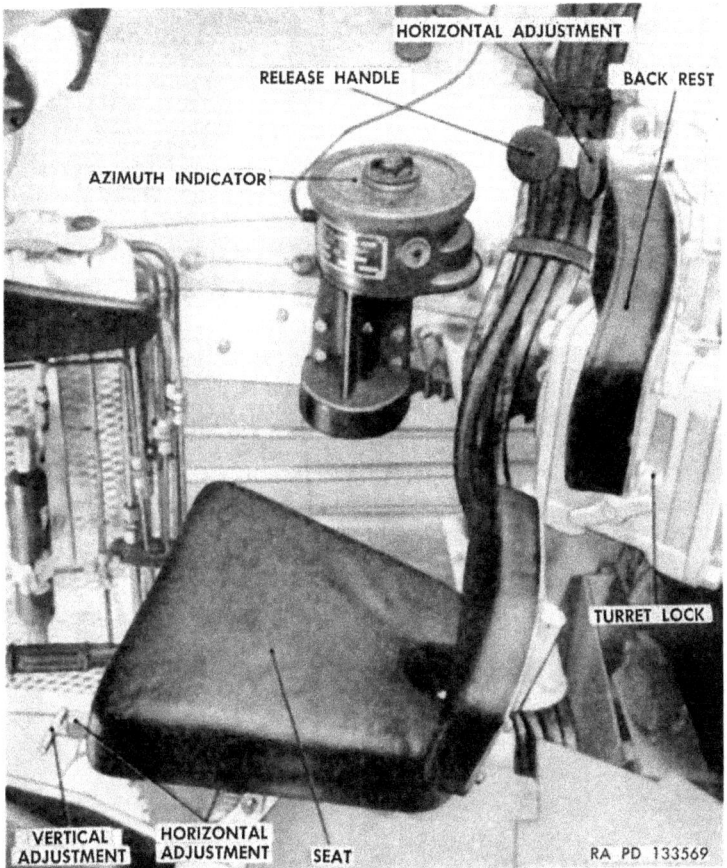

Figure 22. Turret lock.

to avoid excessive wear of the traversing and elevating mechanism due to vibration while the tank is moving.

(2) To open the lock, unsnap the chain from the wing bolt and unscrew the wing bolt from its bracket by turning it counterclockwise. Swing the cap back. Lay the traveling lock back over the rear deck. Before the traveling lock can again be raised to the vertical position, the stop pawl must be released.

(3) To close the lock, swing the cap over the gun tube and tighten it by turning the wing bolt clockwise. Snap the chain to the wing bolt to prevent its rotation.

Section V. TURRET AND ARMAMENT CONTROLS AND EQUIPMENT, M41A1 TANK

27. General

The power control system of the M41A1 provides the same controlled movement of the gun and turret as does that of the M41 tank. In addition, it provides for power elevation of the tank gun by the tank commander. The turret and gun traveling locks are similar to those on the M41 tank (par. 26).

28. Gunner's Power Control, M41A1 Tank

The gunner's power traverse control wheel (fig. 23) is located directly in front of the gunner's position. It provides for power control of the turret by the gunner. The turret can be traversed 360° in either direction as follows: to traverse right, rotate the control wheel clockwise; to traverse left, rotate the control wheel counterclockwise. The amount the wheel is rotated from neutral determines the speed of turret traverse. The maximum speed of turret traverse is four complete revolutions per minute.

29. Tank Commander's Power Control, M41A1 Tank

a. The tank commander's power control handle (fig. 24) is

Figure 23. Gunner's power traverse control wheel, M41A1.

located on the turret roof to the right front of the tank commander's position and gives him the same power control of the turret that the gunner has. In addition, it enables the commander to elevate or depress the tank gun in power.

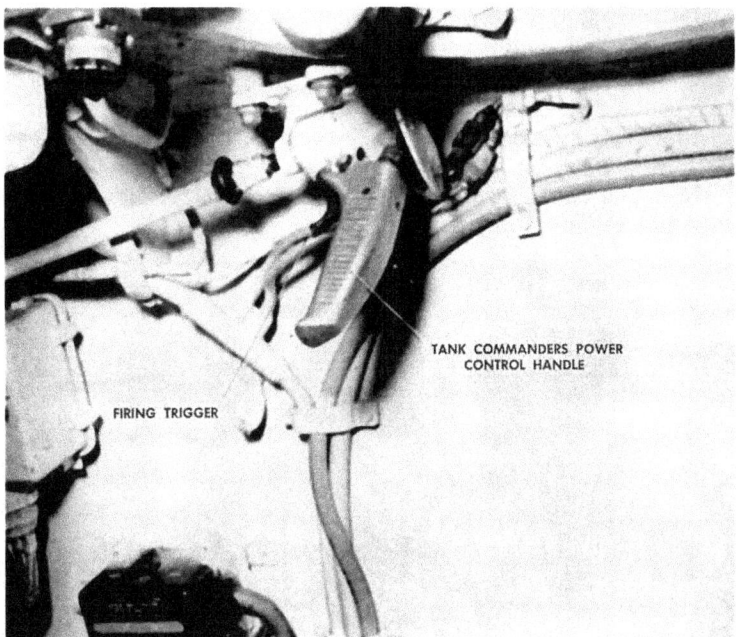

Figure 24. Tank commander's power control handle, M41A1.

b. To traverse the turret, move the butt of the power control handle in the desired direction of travel. The amount the handle is moved from neutral determines the speed of traverse. Before the tank commander can elevate or depress the gun, the elevation switch on the gun control box must be in the ON position. Moving the butt of the power control handle to the rear will cause the gun to be depressed; moving the butt forward will cause the gun to be elevated.

30. Gunner's Manual Controls, M41A1 Tank

a. *Manual Traverse Control Handle.* The gunner's manual traverse control handle (fig. 23) is located above and to the right of the gunner's seat. To traverse the turret manually, grasp the manual traverse control handle, squeeze the release lever on the handle, and rotate the handle in the desired direction of traverse. Several recesses are provided for locking the manual traverse control handle.

A no-back mechanism automatically holds the turret in position and prevents turret drift.

b. Manual Elevation Control Handle. The gunner's manual elevation control handle (fig. 25) is mounted on a handwheel located to the left front of the gunner and on the right side of the gun mount. The handwheel is connected through a differential to a rack-and-gear combination that causes the gun to be elevated or depressed as the handwheel is rotated. To elevate the gun, rotate the handwheel counterclockwise. To depress the gun, rotate the handwheel clockwise. A tension adjusting finger is provided on the elevation handwheel. This finger enables the gunner to adjust the sensitivity or "feel" of the manual elevation control handle to his individual preference.

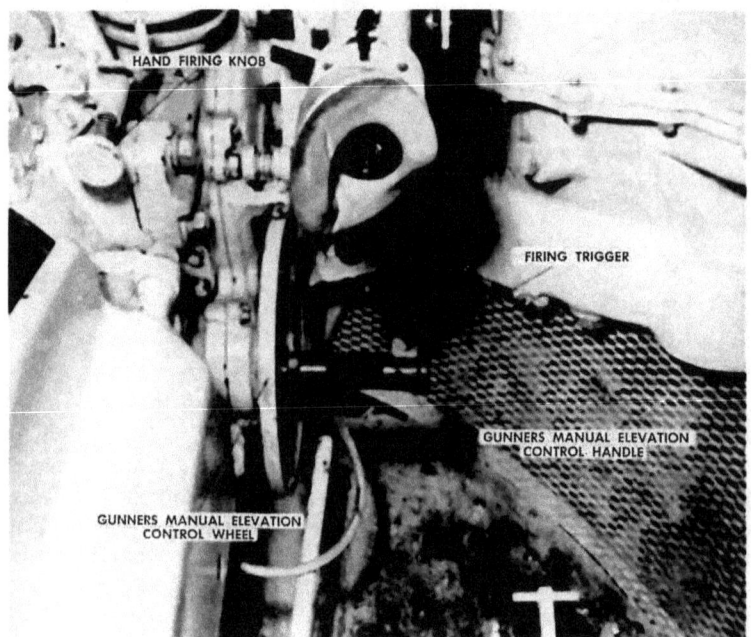

Figure 25. Gunner's manual elevation control wheel and handle, M41A1.

31. Gun Control Box, M41A1 Tank

a. The gun control box (fig. 26) is located on the turret wall to the right of the gunner's position. On it are mounted switches and indicator lights that control power elevation of the gun and the electrical firing circuits.

b. The elevation switch is on the right side of the gun control box. This switch must be in the ON position before the tank com-

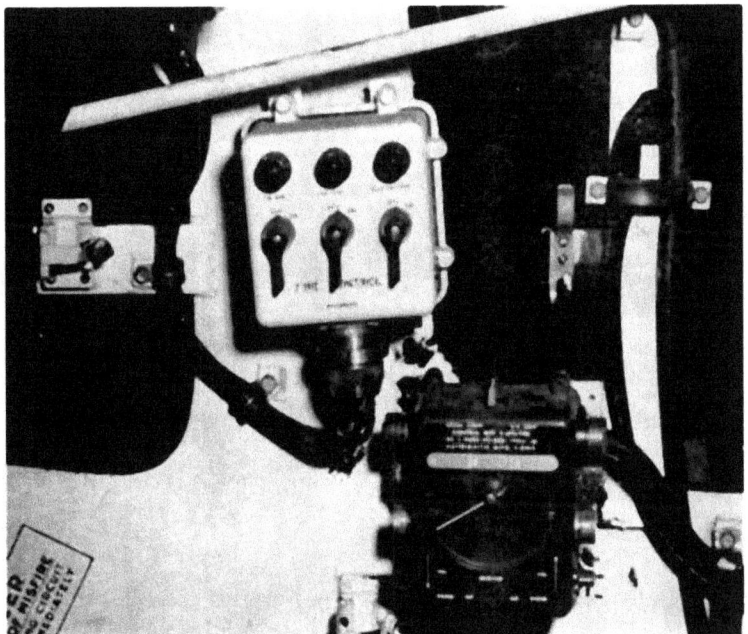

Figure 26. Gun control box, M41A1.

mander can control the elevation or depression of the gun with his power control handle.

c. The coaxial machinegun switch is in the center of the gun control box. This switch controls the electrical circuit for the coaxial machinegun. When the switch is on, the indicator light above it should be on.

d. The 76-mm gun switch is on the left side of the gun control box. This switch controls the electrical firing circuit for the main armament and must be in the ON position before the 76-mm gun can be fired electrically. When the switch is on, the indicator light above it should be on.

32. Turret Motor Switch, M41A1 Tank

The turret motor switch (fig. 27) is mounted on the turret roof above the main armament. This switch must be in the ON position before the turret can be operated in power. This switch is readily accessible to all members of the crew in the fighting compartment.

Note. When the turret motor is running, the noise level is quite high because of characteristics inherent in the system. This is not a cause for alarm if the system has been properly serviced and maintained.

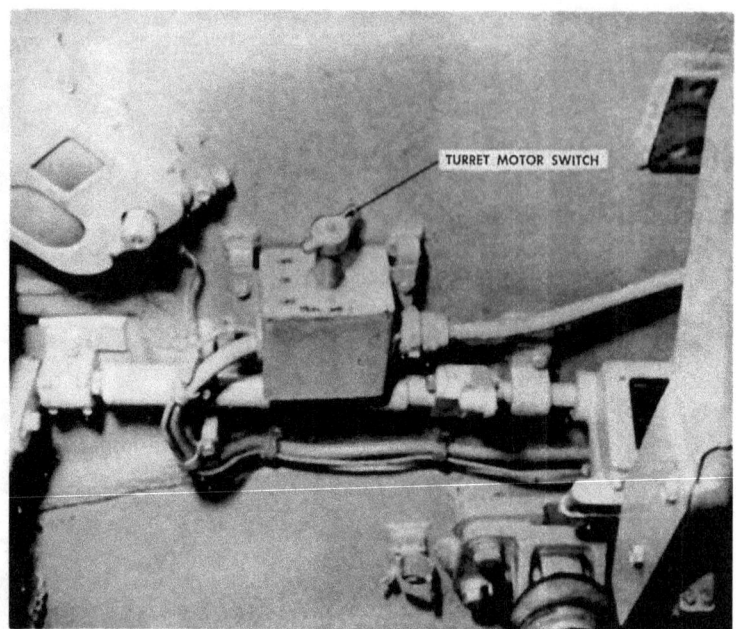

Figure 27. Turret motor switch, M41A1.

33. Firing Controls, M41A1 Tank

a. Electrical. The master relay switch and the appropriate gun switch must be on in order to fire either the 76-mm gun or the coaxial machinegun electrically.

  (1) The gunner may fire the 76-mm or coaxial machinegun electrically by using the firing trigger located on the manual elevation control handle.

  (2) The tank commander may fire the 76-mm gun or the coaxial machinegun electrically by using the firing trigger on the front of the tank commander's power control handle. This trigger will function only when the grip safety is depressed.

b. Manual. No electrical firing circuits are used when the tank gun is fired manually. To manually fire the 76-mm gun, move the hand firing knob forward.

Caution: Release the knob immediately after the gun fires.

c. Safeties.

  (1) *Timing relay (electrical).* When the 76-mm gun is being fired electrically, the timing relay breaks the electrical circuit immediately after the firing plunger is actuated

by the solenoid. A gun-ready light microswitch (fig. 28) is mounted on a bracket to the right of the breech ring. When the breech is opened, the operating crank is rotated away from the switch, causing the gun-ready lights to be turned off. When the breech is closed, the operating crank contacts the switch and the gun-ready lights are turned on.

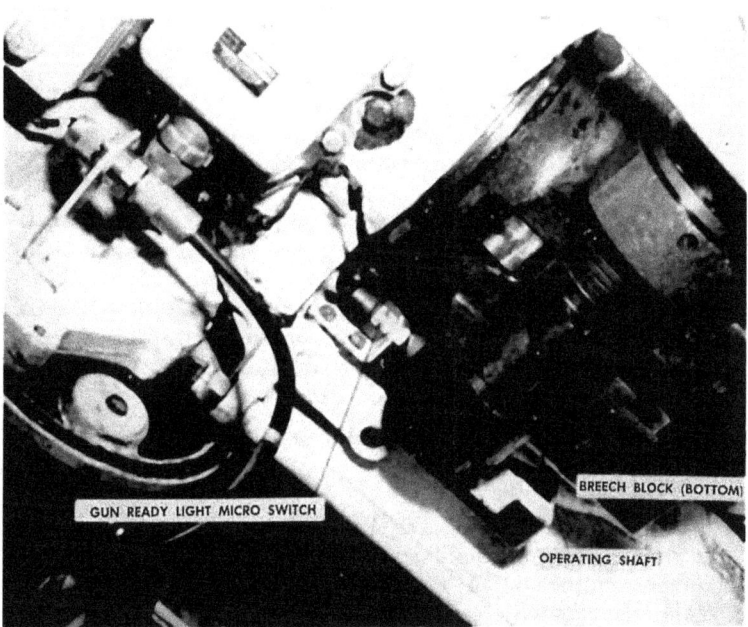

Figure 28. Gun-ready light microswitch, M41A1.

(2) *Manual safety.* The 76-mm gun safety is described in paragraph 25c(2).

34. Steps for Placing Turret Into Power Operation

To prevent injury to crew members or damage to the turret control system and equipment, it is necessary that certain steps be followed when the turret is being placed into power operation. The following steps are performed, using the word "ACUTE" as a guide:

A. *Alert crew*_____Crew members must be in safe positions; check for obstructions on inside and outside of tank.

C. *Check oil level* _____ Oil in reservoir should reach *full* mark on bayonet gauge.

Note. Oil in the traversing and elevating mechanisms of M41A1 tank must be at plug level.

U. *Unlock turret* _____ Turret should be traversed manually to insure that it is unlocked.

T. *Turn on power* _____ Gunner's and commander's power control handles should be in a neutral position. Loader's traverse safety switch must be on. Dump valve toggle switch must be in POWER position.

Note. The elevation switch must be turned on in the M41A1 tank. The loader's traverse safety switch and dump valve toggle switch do not apply to the M41A1 tank.

E. *Elevate and traverse* ___ Check manual and power elevation systems for proper functioning. Note action of power traverse system to determine if it is operating properly.

Note. The M41 tank does not have power elevation.

Section VI. SIGHTING AND FIRE-CONTROL EQUIPMENT

35. Arrangement and Use

a. General. The sighting and fire-control equipment consists of direct-fire equipment, auxiliary fire-control equipment, and miscellaneous equipment. Each crew member has at least one type of vision device available to him.

b. Direct-Fire Equipment. This group consists of—Telescope, M97, mounted in telescope Mount, M92, in the M41 tank or mounted in telescope Mount, M92A1, in the M41A1 tank, with Instrument Light, M36; gunner's Periscope, M20 or M20A1, mounted in periscope Mount M93, with Instrument Light, M36; commander's Periscope, M20 or M20A1, mounted in periscope Mount M94, with Instrument Light, M36; and Ballistic Drive, M4, with Instrument Light, M30. These items are either linked or mounted coaxially with the guns and are used to lay the 76-mm gun and the coaxially mounted machinegun on targets that can be seen from the tank.

c. Auxiliary Fire-Control Equipment. This group consists of an Azimuth Indicator, M31, with an instrument light and Gunner's Quadrant, M1 (M1A1). The M41 tank has an M9 elevation quadrant. These instruments are used to lay the 76-mm gun on targets which are not visible to the gunner.

d. Miscellaneous Equipment. This group consists of four M17 periscopes for the driver; one M19 periscope, which replaces one of the M17 periscopes for night driving; one M13 periscope for the loader; a projection system, which allows the gunner to view the scales of the ballistic unit; and an M27 fuze setter for setting the fuzes of ammunition. Five vision blocks are mounted around the commander's cupola to provide him with all-around vision.

36. Periscope, M20 (M20A1)

a. General. The Periscope, M20 (fig. 29), is a sighting instrument having two built-in optical systems: a one-power system for close-in, wide-angle observation, and a six-power system for sighting. The periscope is composed of two separate parts: a replaceable head assembly which is supported in the top of the periscope mount by two clamping screws, and a body assembly (fig. 30) which is secured in the bottom of the periscope mount with four screws and washers.

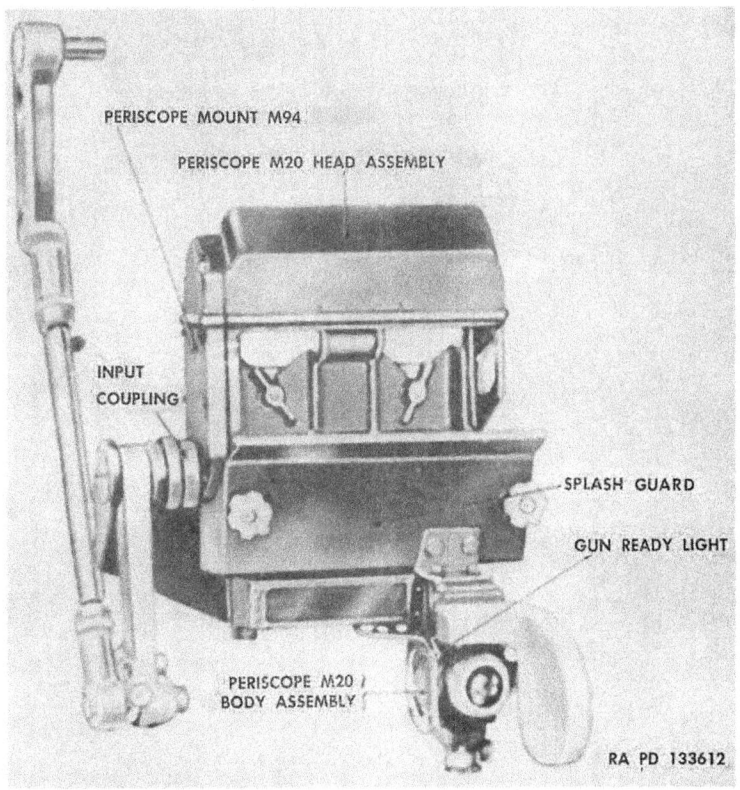

Figure 29. M20 periscope (commander's).

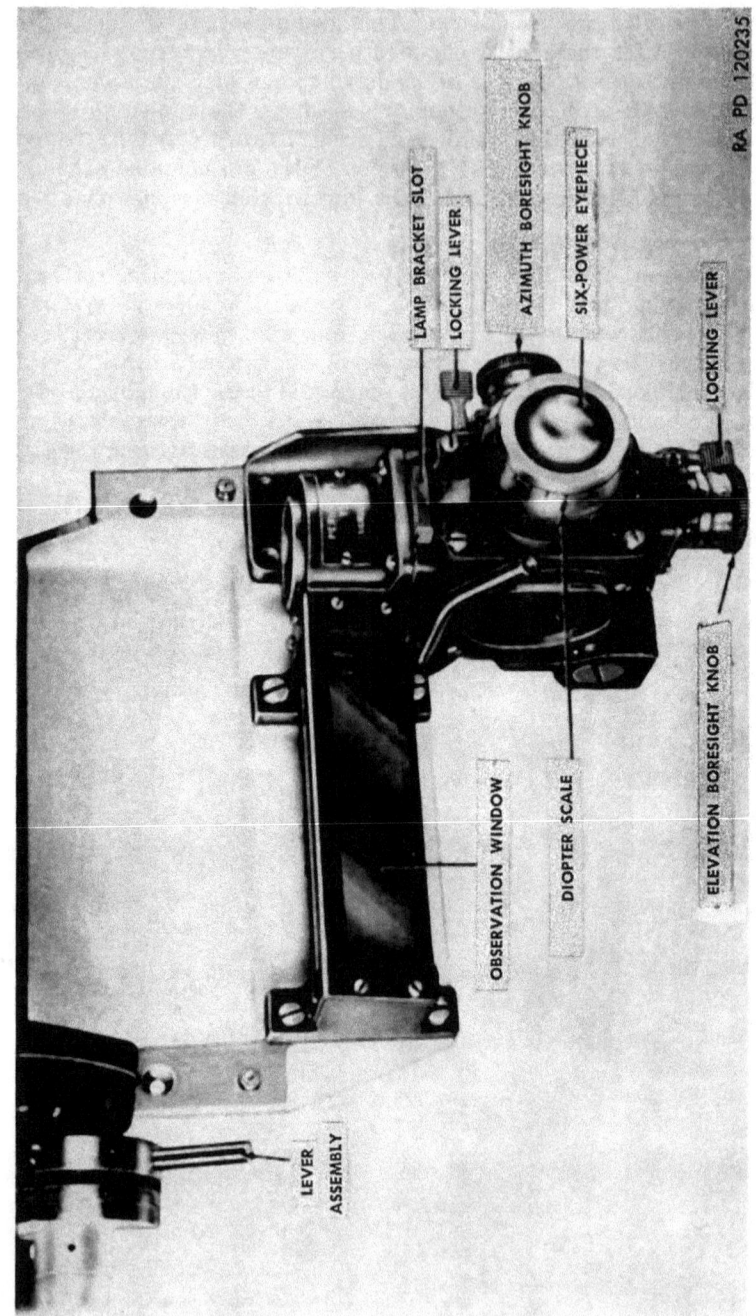

Figure 30. Body, M20 periscope.

b. Arrangement and Use. The gunner's and commander's periscopes are part of the primary direct-fire control system, providing a direct-fire sight for the gunner and a target-designating sight for the commander. The input coupling on the gunner's periscope is coupled directly to the Ballistic Drive, M4 (fig. 33). The input coupling on the commander's periscope (fig. 29) is coupled to an extension arm on the periscope mount which, in turn, is connected to the ballistic drive. Therefore, both periscopes are operated in unison with the gun.

c. Controls.

(1) The line of sight of the periscope can be manually elevated, independent of the gun, 22° from zero elevation and approximately 32° if the gun is at maximum depression. This allows the operator to scan the terrain without elevating the gun. The input coupling is spring-loaded to allow it to return to the original position when released.

(2) A diopter adjustment (fig. 30) is provided on the eyepiece of the six-power system for focusing the periscope to the eye of the user. The diopter scale is graduated from –3 to +3 diopters for presetting the eyepiece when the observer's eye correction is known.

(3) Two boresighting knobs (fig. 30) are provided on the periscope body for making reticle adjustments. Each knob has a locking lever to hold the desired adjustment and also has a scale, graduated from 0.5 to 5.5 mils, for recording the adjustments made.

d. Reticle Pattern. The gun-laying reticle (fig. 31) in the six-power system of the M20 periscope represents an angular measurement of 40 mils in width, with each horizontal line measuring 5 mils. The aiming cross in the center of the reticle, formed by intersecting lines 2 mils long, is used as the boresight point and for laying the gun for the initial round at a stationary target. The one-power system (observation window) of the periscope has no reticle pattern.

e. Illumination. A dovetailed slot and a window are provided above the eyepiece for attaching the lamp bracket of Instrument Light, M36 (fig. 32), for illumination of the reticle. One instrument light is provided for each periscope.

37. Periscope Mounts, M93 (Gunner's) and M94 (Commander's)

a. General. The periscope Mounts, M93 and M94, provide supports for the two M20 periscopes used by the gunner and commander. A guard is installed on the back of each mount to pro-

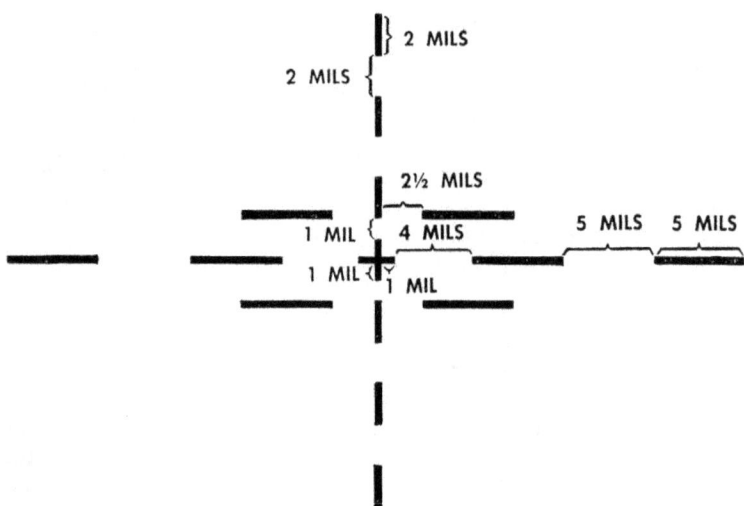

Figure 31. Gun-laying reticle, M20 periscope.

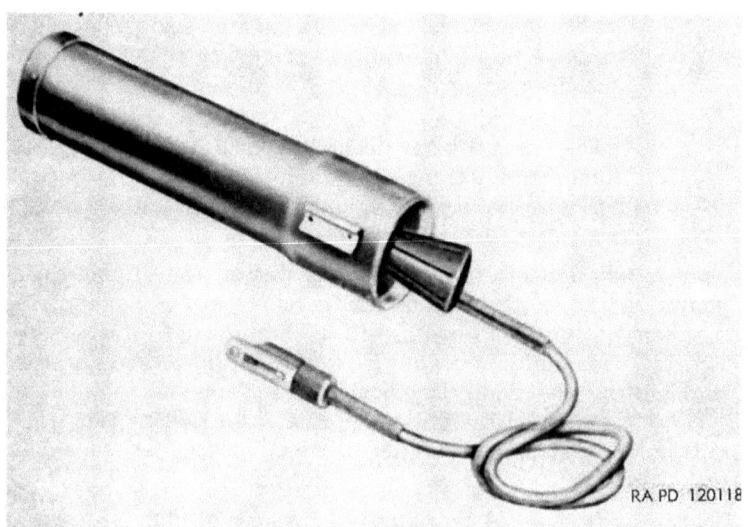

Figure 32. Instrument Light, M36.

tect the operator in the event the periscope head is hit by shell fragments or small-arms projectiles. Locating pins are provided to correctly position the periscope head assembly. Two retaining screws are provided to hold the head asembly in position and to expedite replacement. A level assembly is provided on the M94 mount to aid in synchronization. Each mount is provided with a headrest.

b. Arangement and Use. The M93 and M94 periscope mounts are part of the primary direct-fire control system. The gunner's periscope mount is attached to the turret roof. The commander's periscope mount is attached to the front of the commander's cupola. Linkage on the left side of periscope Mount, M94, connects the input coupling of the periscope to the Ballistic Drive, M4.

c. Illumination. A cable from the loader's reset safety on the M41 tank is connected to a receptacle at the base of periscope Mount, M93, and at the rear of periscope Mount, M94, and supplies power to light a lamp when the loader operates the "gun-ready" signal. An illuminating tube carries the light to the gunner's or commander's position, where it functions as a gun-ready light.

38. Ballistics Drive, M4

a. General. The Ballistic Drive, M4 (fig. 33), consists of a ballistic unit and an arm assembly. The ballistic unit contains a range drum on which there are range scales graduated for various types of ammunition. A mil scale is also provided; this can be used in conjunction with a firing table to index elevations for ammunition which is not included on the ammunition scales. A knob is provided on the ballistic unit to index the desired setting on either a range scale or the mil scale.

b. Arrangement and Use. The ballistic drive provides a means of introducing into the line of sight correct superelevation for the range and type of ammunition being used. Brackets are provided for mounting the ballistic drive to the turret roof adjacent to the gunner's periscope mount. As the range drum is manually rotated to the desired range, the line of sight of each periscope is raised or lowered by means of the connections described in paragraph 36*b*.

c. Illumination. Two built-in electric lamp assemblies are provided for illuminating the scales of the ballistic unit. A twin receptacle is provided at the left rear of the ballistic unit. The 24-volt supply from the vehicle power source is connected to the upper receptacle, while the lower receptacle is provided for receiving the plug from Instrument Light, M30 (fig. 34), which is used for emergency power should the 24-volt supply fail. When the switch is in the lower position, the 24-volt supply is connected to the ballistic unit; in the upper position, the instrument light is connected in the circuit; and in the middle position, no illumination is provided, as both light sources are cut out.

39. Telescope, M97

a. General. The Telescope, M97 (fig. 35), is a straight-tube,

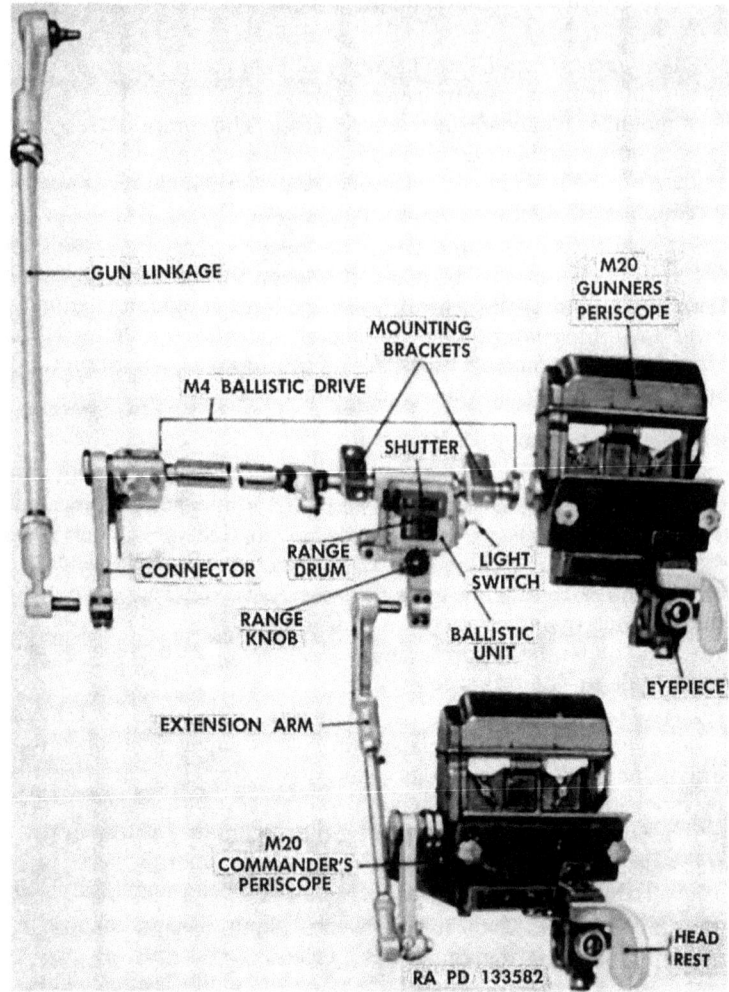

Figure 33. Arrangement of Ballistic Drive, M4, with primary direct-fire equipment.

direct-sighting telescope with a magnification of 8-power and a field of view of 7° 24′. It is provided with a rubber eyeguard, mounted on a bracket, which can be adjusted to vary the distance of the observer's eye from the eyepiece. It is also provided with a diopter scale for focusing to the observer's eye correction.

b. Arrangement and Use. The M97 telescope is part of the secondary direct-fire control system, providing an emergency direct-fire sight for the gunner should the primary system fail. It

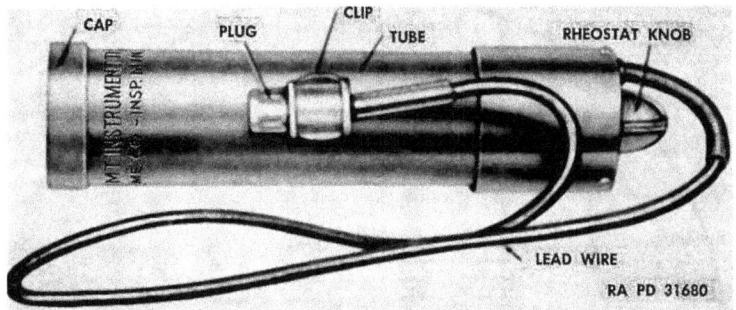

Figure 34. Instrument Light, M30.

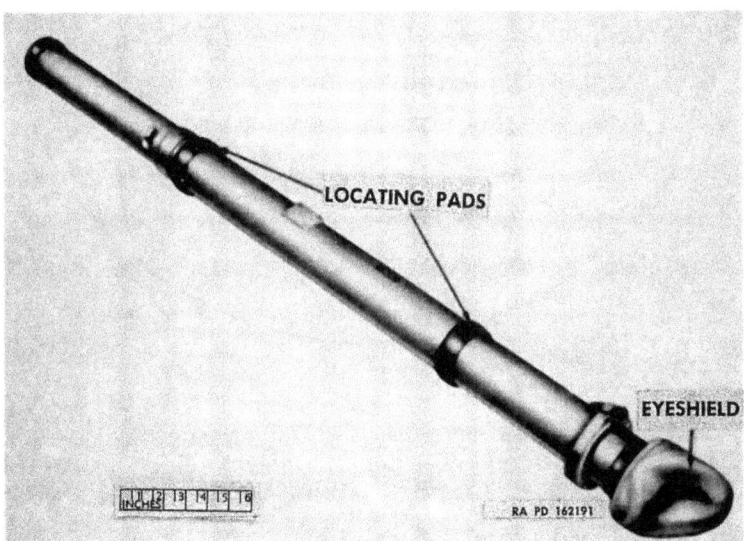

Figure 35. Telescope, M97.

is supported in telescope Mount, M92, in the M41 tank (M92A1 in the M41A1 tank) by means of two mounting collars on the telescope tube.

c. Reticle Pattern. The telescope is provided with a standard ballistic-type gun-laying reticle (fig. 36), which is graduated for AP-T 128E6 ammunition. The center of the boresight cross represents zero elevation and zero deflection and is used when boresighting. Each horizontal line and each vertical space represents 5 mils deflection or one lead. Each vertical line and space represents a range change of 200 yards. An aiming data chart (fig. 37) is provided for use in connection with other types of ammunition.

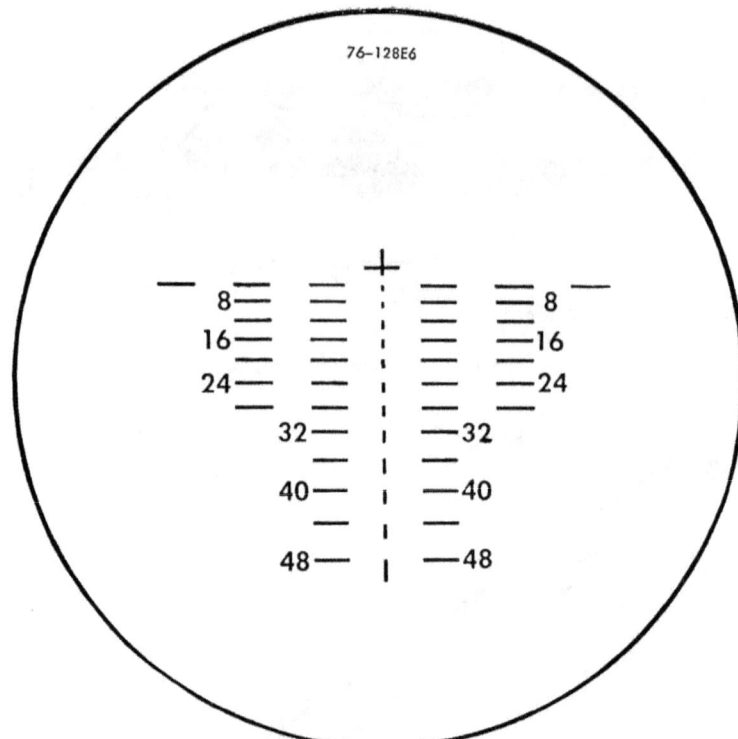

Figure 36. Gun-laying reticle, M97 telescope.

d. Illumination. A dovetailed slot and a window are provided forward of the front collar for attaching the lamp bracket of Instrument Light, M36, when illumination of the reticle is desired.

40. Telescope Mounts, M92 and M92A1

a. General. The telescope Mounts, M92 and M92A1, provide support for the M97 telescope parallel to the 76-mm gun. Each mount consists of a front bracket assembly and mount assembly.

b. Arrangement and Use. The M92 and M92A1 telescope mounts are part of the secondary direct-fire control system, providing a means of mounting the M97 telescope coaxially with the 76-mm gun. The front bracket assembly of each mount is mounted on the gun shield adapter. The mount assembly of the M92 telescope mount is mounted on the top of the 76-mm gun mount in the M41 tank, whereas the mount assembly of the M92A1 telescope mount is mounted to the side of the turret elevating mechanism in the M41A1 tank. The front mounting collar of the M97 telescope is

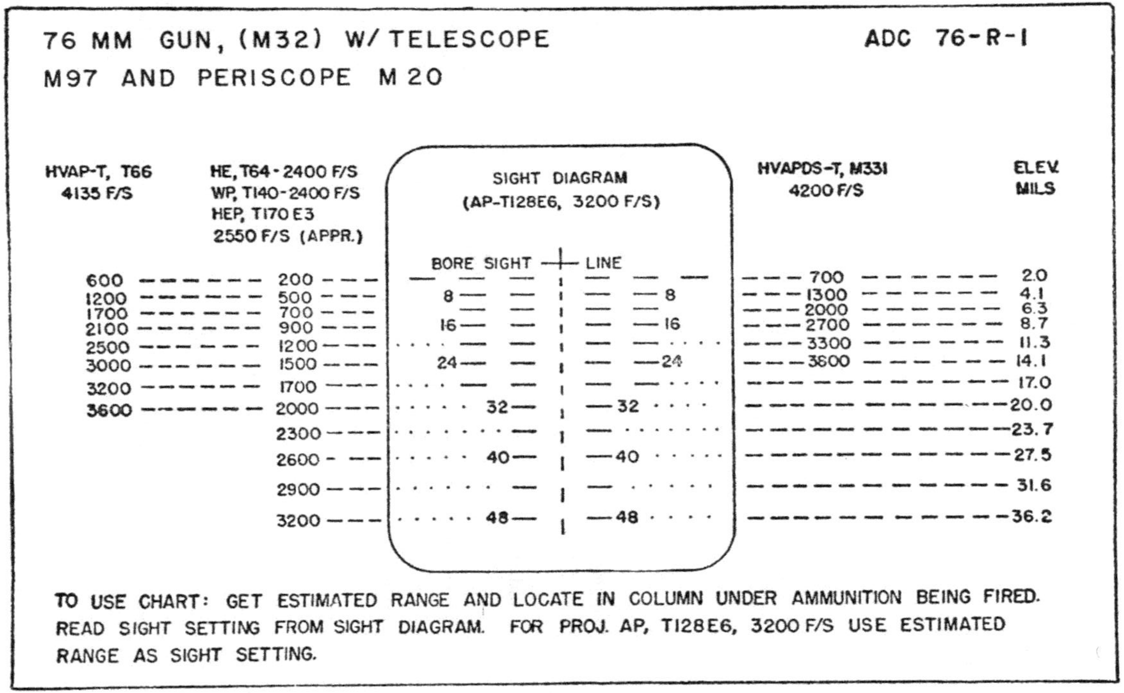

76 MM GUN, (M32) W/TELESCOPE M97 AND PERISCOPE M20

ADC 76-R-1

HVAP-T, T66 4135 F/S	HE, T64 - 2400 F/S WP, T140-2400 F/S HEP, T170 E3 2550 F/S (APPR.)		HVAPDS-T, M331 4200 F/S	ELEV MILS
600 — — — — — —	— — 200 — — —		— — 700 — — — —	2.0
1200 — — — — —	— — 500 — — —		— — 1300 — — — —	4.1
1700 — — — — —	— — 700 — — —		— — 2000 — — — —	6.3
2100 — — — — —	— — 900 — — —		— — 2700 — — — —	8.7
2500 — — — — —	— — 1200 — — —		— — 3300 — — — —	11.3
3000 — — — — —	— — 1500 — — —		— — 3600 — — — —	14.1
3200 — — — — —	— — 1700 — — —		— — — — — —	17.0
3600 — — — — —	— — 2000 — — —		— — — — — —	20.0
	— — 2300 — — —		— — — — — —	23.7
	— — 2600 — — —		— — — — — —	27.5
	— — 2900 — — —		— — — — — —	31.6
	— — 3200 — — —		— — — — — —	36.2

SIGHT DIAGRAM
(AP-T128E6, 3200 F/S)

BORE SIGHT —|— LINE

```
           8 —        — 8
          16 —        — 16
          24 —        — 24
          32 —        — 32
          40 —        — 40
          48 —        — 48
```

TO USE CHART: GET ESTIMATED RANGE AND LOCATE IN COLUMN UNDER AMMUNITION BEING FIRED. READ SIGHT SETTING FROM SIGHT DIAGRAM. FOR PROJ. AP, T128E6, 3200 F/S USE ESTIMATED RANGE AS SIGHT SETTING.

Figure 37. Aiming data chart.

held in the front bracket assembly by a thumbscrew, and the rear mounting collar is held in a holder assembly on the mount assembly. The upper half of the holder assembly is hinged and fastened by a wing nut to permit installation and removal of the telescope.

c. Controls. An azimuth and an elevation adjusting knob are provided to permit adjustment of the telescope during boresighting and zeroing. Each knob is provided with a scale graduated from 0 to 4 mils. A wing nut, which can be loosened to permit the knob to be moved off the serrated shaft, is provided to allow additional adjustment, and in order to reset the knob to the normal (2) position after adjustment. A locking lever is provided on each mechanism to retain the desired setting.

d. Illumination.

(1) A cable from the loader's safety panel is connected to a receptacle on the mount and supplies power to light a lamp when the loader operates the gun-ready signal. An illuminating tube carries the light from the lamp to the gunner's viewing position. This light functions as a gun-ready signal.

(2) The M36 instrument light is supported in a clamp on the telescope mount and is provided to illuminate the reticle of the telescope.

41. Azimuth Indicator, M31

a. General. The Azimuth Indicator, M31 (fig. 38), which is mounted on the right side of the turret with gears in mesh with the turret ring, measures horizontal angles of traverse. It is used principally for laying the gun for indirect fire. The azimuth indicator is a dialed instrument with pointers indicating the readings on the scales.

b. Arrangement and Use.

(1) The Azimuth Indicator, M31, has three scales and three pointers. The azimuth scale is graduated in 100-mil intervals and is numbered every 200 mils from 0 to 3,200 counterclockwise in two consecutive semicircles around the scale. The micrometer scale is graduated counterclockwise in 1-mil intervals and numbered every 5 mils from 0 to 100. The gunner's aid is graduated in 1-mil intervals and numbered every 5 mils from 0 to 50 mils right and left. The directional pointer is fixed in relation to the longitudinal axis of the tank and gives a course reading on the azimuth scale. This reading indicates the number of mils the gun has traversed from the longitudinal axis of the hull. The azimuth pointer works in con-

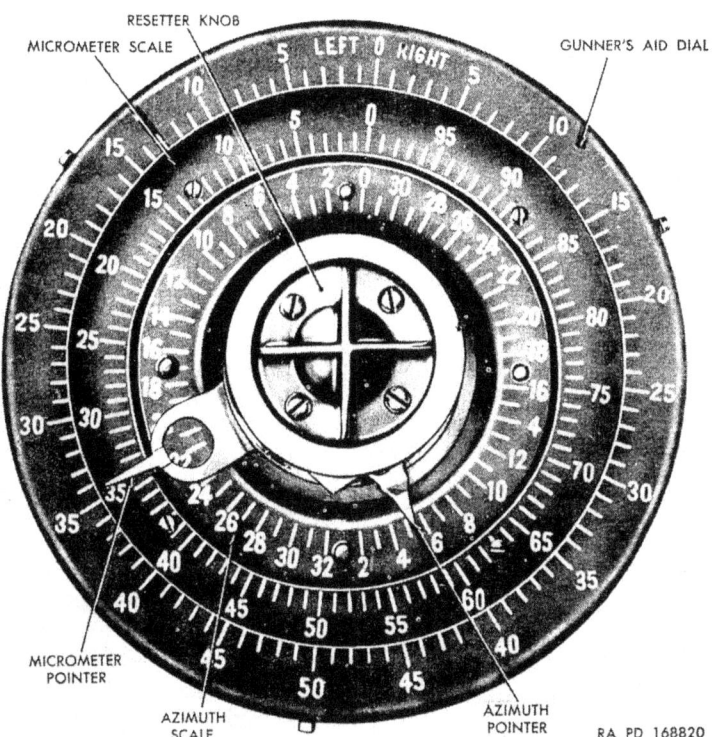

Figure 38. Azimuth Indicator, M31.

junction with the micrometer pointer. These pointers are adjustable and may be set at zero (by depressing and turning the resetter knob) when the sights are set on any desired reference point. The sum of these two pointer readings then gives a precise reading of the number of mils the gun traverses from the reference point. The azimuth and micrometer scales are fixed, while the gunner's aid dial may be rotated to any position. The gunner's aid is used in making deflection corrections by rotating the dial until the zero coincides with the position of the micrometer pointer. Right and left deflection corrections are then laid off on the gunner's aid dial. After the correction has been applied, move the zero of the gunner's aid to the micrometer pointer.

(2) Built-in electric lamps provide illumination for the scales of the azimuth indicator. A receptacle on the side of the azimuth indicator receives the plug on the instrument

light, which is installed in a bracket immediately above the indicator. The lamps, which are in the indicator, are turned on and off by a toggle switch on the instrument light.

(3) To test the accuracy of the azimuth indicator, lay the aiming cross of the periscope on a definite aiming point, and set the azimuth and micrometer pointers at zero. Traverse the turret manually through a complete circle until the sight is laid back on the original aiming point. If the azimuth and micrometer pointers do not indicate zero, the azimuth indicator is out of adjustment and needs a check by ordnance personnel.

(4) To test the azimuth indicator for slippage, lay the aiming cross of the sight on a definite aiming point, and set the azimuth and micrometer pointers at zero. Traverse the turret rapidly in power, and stop suddenly; repeat this operation two or more times in the same direction. Manually traverse the turret back to the aiming point. If the azimuth and micrometer pointers do not indicate zero, the azimuth indicator is slipping and will require adjustment by qualified ordnance personnel. If the pointers indicate zero, repeat the procedure in the opposite direction.

(5) To maintain the azimuth indicator, keep it clean and covered when not in use. Any lubrication or adjustment must be done by ordnance personnel.

42. Gunner's Quadrant, M1 (M1A1)

a. Description.

(1) The Gunner's Quadrant, M1 (fig. 39), consists of a sector-shaped frame to which is pivoted an index arm with a level holder. A scale graduated from 0 to 800 mils is on one side of the frame, and a scale graduated from 800 to 1,600 mils is on the opposite side. Both scales are graduated in 10-mil intervals and numbered every 50 mils.

(2) A level and micrometer mechanism is mounted on the index arm. The micrometer has two scales graduated in .2-mil increments from 0 to 10 mils, and numbered every mil. On the M1 quadrant, one scale is in black and one in red. The black figures are for readings of 0–800 mils. The red figures are not used in tank gunnery. On the M1A1 quadrant, one scale is visible only when reading the 0–800-mil side of the quadrant and the other scale is visible only when reading the 800–1,600-mil side.

(3) Two sets of bearing shoes are provided, displaced 90°, which serve as bearing surfaces when the quadrant is

being used with either scale. An arrow is provided on each side of the frame to indicate the direction in which the quadrant is to be faced when each scale is being used.

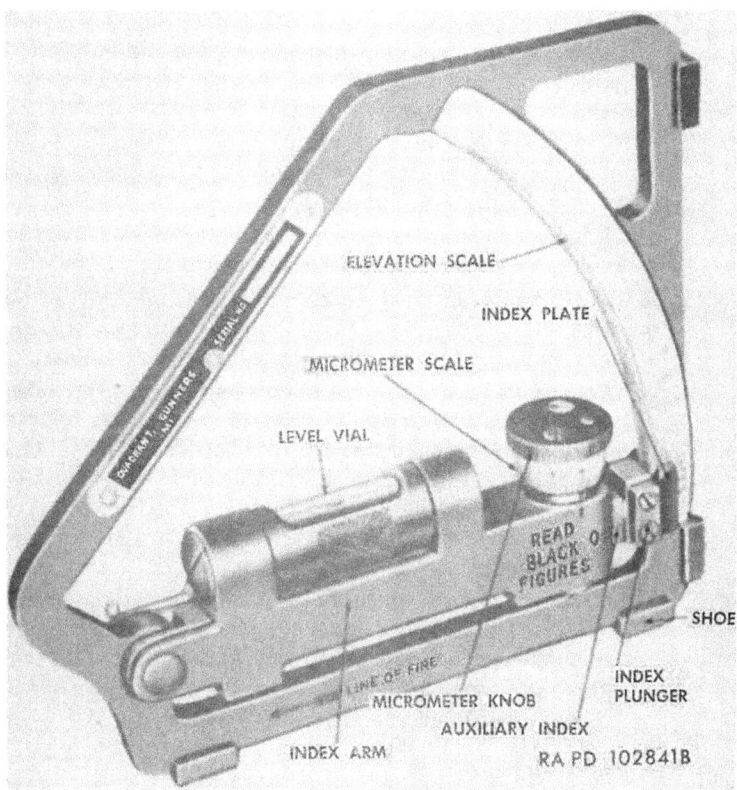

Figure 39. Gunner's Quadrant, M1.

b. *Arrangement and Use.*

(1) The M1 gunner's quadrant is provided as part of the indirect-fire control system and is used in conjunction with the M31 azimuth indicator when the gun is being laid on a target which is not visible from the tank.

(2) When used, the quadrant provides a means of applying the necessary elevation setting to the gun. The required elevation is set on the elevation scale and micrometer, the quadrant is placed on the pads provided on the breech of the gun, and the gun is then elevated until the level vial in the quadrant indicates *level*.

(3) When not in use, the quadrant is placed in the carrying case provided and is stored in the bustle of the tank.

c. *Tests and Adjustments.* To check the accuracy of the M1 gunner's quadrant, proceed as follows:
   (1) Set the index arm and micrometer at zero, and aline the index marks.
   (2) Place the quadrant on the pads provided on the breech of the gun, using the 0–800-mil bearing shoes, and center the level bubble by elevating or depressing the gun.
   (3) Turn the quadrant end for end on the breech. If the level bubble recenters itself, the quadrant is in perfect alinement at zero elevation.
   (4) If the level bubble does not recenter itself, attempt to center it by turning the micrometer knob.
   (5) If the bubble can be recentered with the micrometer knob, the correction is plus. It is one-half the reading on the micrometer scale and must be added to all settings.
   (6) If the bubble cannot be recentered by turning the micrometer knob, set the index arm at −10 mils on the elevation scale (one notch below zero). Then center the bubble with the micrometer knob. In this case, the correction is minus and must be subtracted from all settings. Subtract the setting on the micrometer scale from 10, and divide the remainder by 2.
   (7) If the correction exceeds .4-mil, send the quadrant to ordnance for adjustment. Using units are not permitted to make any adjustment of the quadrant. In an emergency, an out-of-adjustment quadrant can still be used by applying the correction to all readings.

43. Elevation Quadrant, M9

a. *General.* The Elevation Quadrant, M9 (fig. 40), is mounted on the M41 tank on a bracket that is bolted to the right side of the gun mount. It is used to lay the gun for elevation and to measure the elevation of the gun.

b. *Description.* The Elevation Quadrant, M9, consists of—
   (1) Two elevation scales and two elevation scale indexes, one on each side of the elevation quadrant, graduated from −200 to +600 mils.
   (2) An elevation micrometer scale graduated from 0 to 100 mils in units of 1 mil.
   (3) Two indexes for the elevation micrometer scale: one on the loader's side of the gun, and the other on the gunner's side.
   (4) A level vial with a movable cover.

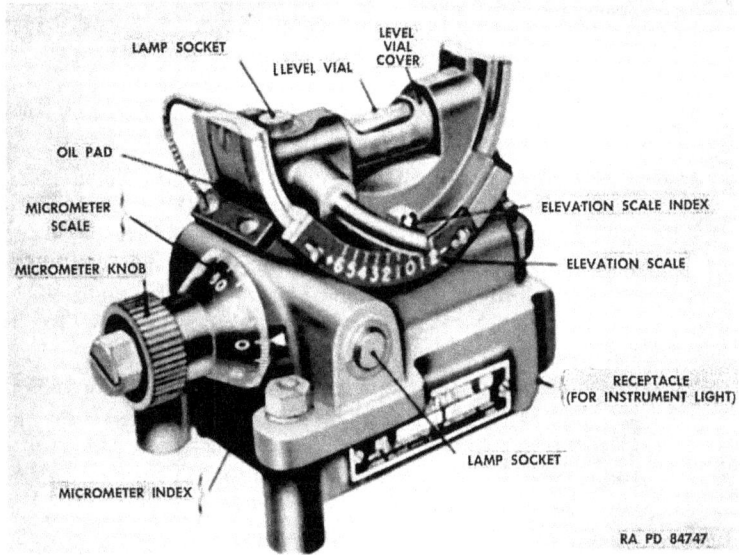

Figure 40. Elevation Quadrant, M9.

(5) A night lighting device for the level vial, elevation scales and indexes, and elevation micrometer and indexes.

c. *Adjustment.*
  (1) Select a gunner's quadrant that is accurate; set it at zero. Using the gunner's quadrant, lay the gun at zero elevation.
  (2) Level the elevation quadrant bubble, using the micrometer knob. The elevation and micrometer scales on the gunner's side should read zero. (The elevation scale on the loader's side should also read zero.)
  (3) If the elevation scales do not read zero, loosen the screws on each end of the elevation scales and slide the scales until the zero graduations are opposite their indexes. Tighten and check them.
  (4) If the micrometer scale does not read zero on the gunner's side, loosen the nut holding the micrometer knob; without turning the knob, rotate the micrometer scale until the zero graduation is opposite the index on the gunner's side. Tighten and check it.
  (5) Only ordnance personnel are authorized to make further adjustments.

d. *Laying the Gun With the M9 Elevation Quadrant.*
  (1) As an example, to set off an elevation of 135 mils, turn

the micrometer knob until the elevation index reaches the "1" on the plus side of the elevation scale. Then turn the micrometer knob counterclockwise until a reading of 35 is opposite the micrometer index on the gunner's side of the quadrant. Elevate or depress the gun until the bubble is accurately centered.

(2) To set off an elevation of minus 135 mils, turn the micrometer knob until the index reaches the "1" on the minus side of the elevation scale. Next subtract the tens and units (35) from 100; the difference is 65. Now set that off on the micrometer scale by turning the knob clockwise. Elevate or depress the gun until the bubble is centered.

(3) For making a small change from a previous setting, count the tens and units as they pass the index. When the micrometer knob is turned counterclockwise (fig. 40), the elevation is increased; when it is turned clockwise, the elevation is decreased.

e. Measuring Elevation With M9 Elevation Quadrant.
(1) Level the bubble with the micrometer knob.
(2) Read the gun elevation from the elevation and micrometer scales. If the elevation scale index is on the minus side of zero, subtract the micrometer reading from 100 to obtain the tens and units of the minus elevation.

f. Care and Maintenance of the M9 Elevation Quadrant.
(1) Keep the bubble covered when not in use.
(2) Protect the scales from damage.
(3) Be careful not to hit the elevation quadrant; do not force the moving parts.
(4) Keep the quadrant clean and free from dust.
(5) Keep the felt oil pads at the ends of the moon-shaped segment pliable; apply a few drops of oil to them occasionally.

44. Periscope, M17

The driver's position is supplied with four nonadjustable, M17 periscopes (fig. 41). These are mounted to provide approximately 200° observation to the driver's front, right and left hand, and left rear. Each periscope is held in its mount by means of two periscope retaining levers which permit removal and replacement of damaged periscopes. A stowage box, providing space for two spare periscopes, is positioned forward of the driver's seat on the wall of the hull.

Figure 41. M17 periscope.

45. Periscope, M13

The loader's position is provided with one adjustable M13 periscope (fig. 42) mounted in the left side of the turret roof. The mount permits elevating or depressing the line of sight but does not permit traversing to either side. The periscope is held in position, within the mount, by a locking knob on the back of the periscope. A lever on the rear portion of the mount prevents the periscope from dropping out of the mount if the locking knob should accidentally vibrate loose or if the loader should desire to partially lower the device when it is not in use. A rubber seal, mounted around the edge of the opening where the periscope extends out of the mount, prevents water and dust from entering the turret.

Figure 42. M13 periscope.

46. Periscope, M19

a. Description.

(1) The Periscope, M19, is an infrared viewing device of the binocular type, used for night driving. Invisible infrared rays are projected forward from the infrared headlamps at the bow of the vehicle. The periscope converts the infrared image to a visible image which is viewed through conventional lenses.

(2) A headrest is provided for the comfort of the operator. A headrest clamping screw on each side of the bracket permits the headrest to be adjusted to suit the observer.

b. Arrangement and Use.

(1) When night driving is necessary, the M19 periscope is installed in its mount in the driver's compartment and the high-voltage cable is connected. With the master relay switch in the ON position, the main light switch set to BO DRIVE, and the IR–BOD switch set to IR, the periscope is energized by turning on the blackout receiver switch. A warmup period of approximately one minute is required before the periscope can be operated.

(2) When not in use, the M19 periscope is stored in a stowage box provided in the driver's compartment, and the high-voltage cable is connected to a dummy receptacle.

c. Controls.

(1) An elevation setting lever, secured to the right side of the periscope body, provides for locking the periscope at the desired elevation. In use, the elevation setting lever is normally adjusted to provide resistance which the operator can overcome by pressing forward into the headrest.

(2) A left and a right focus control, located on the bottom of the periscope body, provide for external adjustment of the resistors for electrostatic focus.

47. Vision Blocks

Five vision blocks are provided in balanced positions around the commander's cupola. These five vision blocks and the commander's periscope provide all-around vision at all times.

48. Projection System

The projection system consists of a lower reflector assembly, which is mounted to the side of the gunner's periscope, and an upper reflector assembly, which is mounted to the turret roof above the ballistic drive. It is provided to enable the gunner to view the scales of the ballistic drive without changing his position.

49. Fuze Setter, M27

Fuze Setter, M27, is a wrench-type fuze setter, which fits over the nose of mechanical time fuzes and allows the operator to turn them to the desired setting, as read on the scale of the fuze.

50. Boresighting and Zeroing

a. General. The purpose of boresighting is to establish a fixed relationship between the gun and the sights. Boresighting is done at a range as near 1,500 yards as possible; when this is done, the line of sight through the gun tube and the direct-fire sights will converge on the boresight point at 1,500 yards. The gun-and-sight combination is then zeroed at this range to insure that the line of sight and the point of impact of the projectile coincide. Boresighting and zeroing are accomplished as outlined below.

b. Boresighting.

  (1) Position the tank as nearly level as possible.
  (2) Open the breechblock, and insert the breech boresight in the chamber. If the breech boresight is not available, remove the percussion mechanism and, with the breechblock closed, use the firing pin well as the breech boresight. Attach the muzzle boresight to the muzzle; or, if the muzzle boresight is not available, tape black thread to form a cross over the witness lines on the muzzle attachment.
  (3) Select a target at a range as near 1,500 yards as possible (preferably a target with clearly defined horizontal and vertical lines).
  (4) Position the right telescope of a binocular over the firing pin well. Using the firing pin well (breech boresight) as a rear sight, and the cross threads on the muzzle as a front sight, aline the axis of the bore on the boresight point by use of the manual traverse and elevation controls (fig. 43). Always make the final lay of the gun from the same direction, for both elevation and deflection.

  (a) *Periscope, M20.*
   1. Set the range scale in the ballistic unit of the ballistic drive at zero by turning the range knob counterclockwise until it stops.
   2. Sighting through the commander's and gunner's periscope eyepieces, aline the aiming cross of each reticle exactly on the boresight point. Make this alinement by rotating the elevation and deflection boresight knobs without disturbing the lay of the gun.
   3. Clamp the boresight knobs with the locking levers, and slip each scale to read "3."

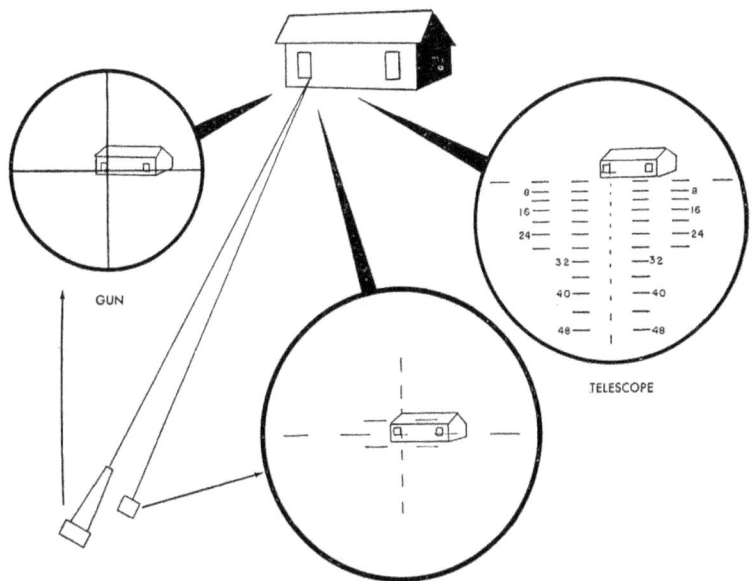

Figure 43. Boresighting.

(b) *Telescope, M97.*
1. Sighting through the telescope eyepiece, aline the boresight cross exactly on the boresight point. Make this alinement by unlocking the knobs with the locking levers, then rotating the elevation and deflection adjusting knobs without disturbing the lay of the gun.
2. Lock the knobs with the locking levers. Loosen the wing nuts, slide each knob off the serrated shaft, and turn each knob until "2" is indicated on the scale. Slide the knobs back on the shafts and tighten the wing nuts.

(c) Recheck all boresight settings.

c. *Zeroing.*
(1) Boresight (b, above).
(2) Select a target at the same range as the point selected for boresighting.
(3) Turn the range knob until the range scale indicates the known range to the target for the type of ammunition to be fired. (Shot should be used.)
(4) Using the traversing and elevating controls, place the aiming cross of the gunner's periscope on a clearly defined aiming point on the target (fig. 44).
(5) Fire from three to five rounds to obtain a shot group. Check and correct the lay of the gun after each round (fig. 44).

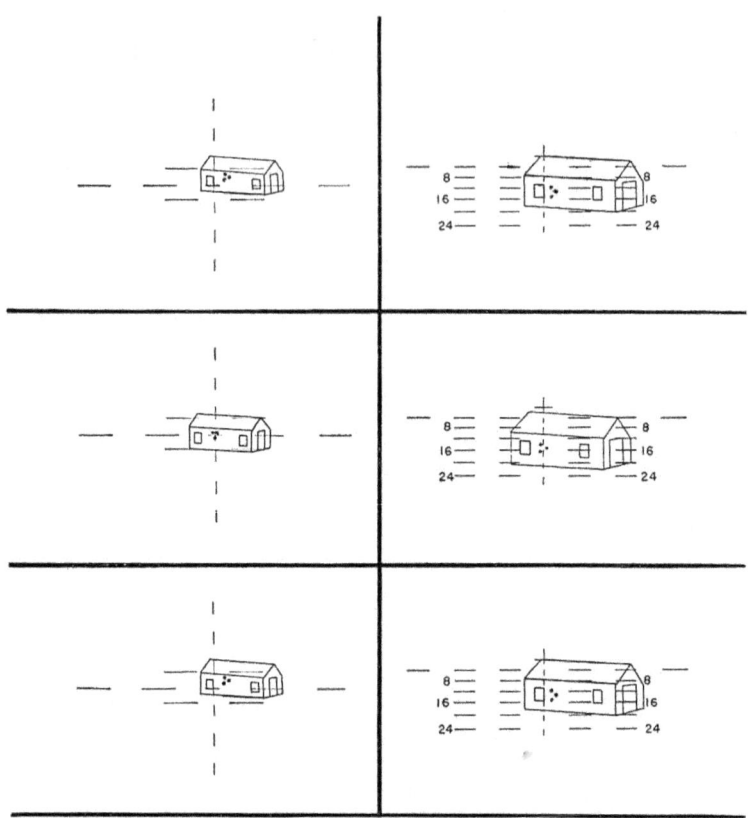

Figure 44. Zeroing.

(6) Without disturbing the lay of the gun, and using the elevation and deflection boresight knobs, lay the aiming cross on the center of the shot group. Clamp the boresight knobs (fig. 44).

(7) Repeat (6) with the tank commander's periscope.

(8) Re-lay on the aiming point of the zeroing target with the manual elevation and traverse controls, and fire a check round. (The shot should hit within 14 inches of the aiming point.)

(9) Re-lay on the aiming point, unlock the boresight knobs of the M97 telescope, and move the reticle until the appropriate range line, corresponding to the known range, is laid on the aiming point (fig. 44). Lock the boresight knobs.

(10) Record all boresight knob settings in some convenient

place in the tank turret and, in pencil, in the gun book. In subsequent sight adjustment it is necessary only to boresight as outlined in *b* above, unlock the boresight knobs, index the current zero settings, and lock the knobs.

Note. For realization of first-round hit potential, great care must be taken in boresighting and zeroing. Check rounds must be fired periodically at definitely known ranges, and rezeroing must be accomplished when required. The accurate recording and use of all zeroing and check firing data in the gun book is essential to first-round hits as well as to conservation of ammunition.

51. Synchronization and Backlash

a. Synchronization. The linkage arm between the gun trunnion and periscope must be adjusted so that the M20 periscope accurately reproduces the movement of the gun in elevation. Check synchronization as follows:

(1) Boresight on a distant aiming point.
(2) Place the tank on a steep forward slope, and check the boresight when the gun is near maximum elevation.
(3) Place the tank on a steep reverse slope (at the same range) and check the boresight when the gun is near its maximum depression.
(4) If the periscope does not check accurately within .5 mil with the gun at maximum elevation or depression, the linkage should be adjusted.

Note. The actual adjustment of linkage must be made by ordnance personnel.

b. Backlash. A backlash check can be performed as follows:

(1) With the gun elevated above a distant aiming point, depress the gun until the boresight cross of the periscope is on the aiming point. Carefully measure the existing elevation with the gunner's quadrant.
(2) Depress the gun below the aiming point, and then elevate it until the aiming cross is again alined precisely on the aiming point.
(3) Measure the existing elevation with the gunner's quadrant.
(4) The difference between the two quadrant readings is the backlash in the system. If it exceeds .3 mil, notify ordnance personnel.
(5) Check backlash of both the gunner's and the commander's periscopes.

Note. The effects of backlash can be largely eliminated if the gunner makes the final lay of the gun from the same direction each time.

CHAPTER 3

CREW DRILL AND SERVICE OF THE PIECE

Section I. GENERAL

52. General

This chapter is for the guidance of platoon leaders and tank commanders in training crew members to attain efficient teamwork in the operation of the M41 or M41A1 tank. It is emphasized that the drills described in this chapter are for the development of crew teamwork in the fighting operation of the tank and that the ultimate goal is successful operation of the tank on the battlefield.

Section II. CREW COMPOSITIONS AND FORMATIONS

53. Crew Composition

The crew of the M41 or M41A1 tank consists of four members: tank commander, gunner, driver, and loader.

54. Formations

a. Dismounted Posts. The crew form in one rank with the tank commander two yards in front of the right track. The gunner, driver, and loader take posts on line with, and to the left of, the tank commander at close interval.

b. Mounted Posts. The crew form mounted as follows:
   (1) *Tank commander.* In the turret, standing on the turret floor platform or seated on the tank commander's seat.
   (2) *Gunner.* In the gunner's seat on the right side of the tank gun and in front of the tank commander.
   (3) *Driver.* In the driver's seat in the left front of the hull.
   (4) *Loader.* On the left side of the tank gun, standing on the turret floor or seated on the loader's seat at the left rear of the turret.

Section III. CREW CONTROL

55. Operation of Interphone and Radio

The tank interphone system is used for voice communication among members of the tank crew and for communication with individuals outside the tank through the external interphone. The tank radio set is used for communication with other tanks and with other units. The interphone is a part of the vehicular radio set. The equipment is designed so that operation of the interphone system will override received or transmitted signals but will not cut transmitted signals off the air. The crew must be proficient in the operation of the interphone system if they are to obtain its maximum value in combat. Proficiency in the operation of the interphone system is gained only by continued practice.

56. Control Box Positions

a. Interphone control box positions are as follows:

(1) The *tank commander* and *gunner* plug into a single control box located on the right wall of the turret.

(2) The *loader* plugs into a control box on the left wall of the turret.

(3) The *driver* plugs into a control box located to his right.

b. External interphone control box positions are as follows:

(1) The external interphone OFF-ON switch is installed in the driver's compartment.

(2) The external interphone PUSH-TO-SIGNAL control is installed in the vehicle turret in the vicinity of the loader's position.

(3) A cable reel, which holds 40 feet of cable terminating with a handset, is located in a compartment at the rear of the tank hull.

57. Modes of Operation

a. General. When power has been supplied to set 1, set 2, and the auxiliary receiver, and after squelch adjustments have been made, the following modes of operation are possible at each interphone control box:

(1) Monitoring of set 1, set 2, and the auxiliary receiver.

(2) Push-to-talk operation of set 1 or set 2.

(3) Interphone facilities between interphone boxes.

b. Monitoring. Monitoring of received signals is accomplished by placing the selector switch pointer of the interphone control box in the center position. This position permits monitoring set 1, set 2, and the interphone in the Radio Set, AN/GRC-4, -6, and -8. It also permits monitoring the auxiliary receiver in the Radio Set, AN/GRC-3, -5, and -7.

c. Interphone Operation. Interphone reception is possible with the selector switch in any position. To communicate with a crew member at any interphone box, press either the HOLD ON or the LOCK ON button on the chest set and talk into the microphone. In an emergency, any crew member can override a radio conversation without waiting for the sending party to stop talking.

d. External Interphone Operation. To contact troops outside the tank, the driver turns on the external interphone; the loader operates the external interphone PUSH-TO-SIGNAL switch, thereby illuminating the external call light located on the bottom of the external interphone compartment. This light can be blinked by momentary operation of the switch. Troops outside the tank, responding to this signal, talk to the crew by removing the handset from the compartment, pressing the PUSH-TO-TALK-AND-LISTEN switch, and speaking into the handset over the tank interphone system. Volume level may be adjusted by turning the volume control in the external interphone compartment. Upon completion of the conversation, the handset and any unreeled cable is returned to the compartment, and its lid is closed and latched. To contact the tank crew from the outside interphone position, with the EXTERNAL-INTERPHONE ON-OFF switch in the ON position, remove the handset from the rear compartment and press the control switch. This illuminates the interphone control light, and communication with the crew can be established.

e. Radio Operation of Sets 1 and 2.
(1) For push-to-talk operation of set 1, turn the selector switch pointer to the left-hand position, press the HOLD ON button and the RADIO button on the chest set, place the RADIO TRANS switch on the control box in the TRANS position, and talk into the microphone. Release the chest set buttons to listen. If the auxiliary receiver is not being used as a monitor station and it interferes with operation of set 1, turn the receiver VOLUME control to the OFF position.
(2) For push-to-talk operation of set 2, turn the selector switch pointer to the right-hand position, press the HOLD ON button and the RADIO button on the chest set, place the RADIO TRANS switch on the control box in the TRANS position, and talk into the microphone. Release the chest set buttons to listen. When set 2 is used, the loader is designated as monitor-operator.

58. Radio Check

Inspection of communication equipment will be performed as prescribed on DA Form 11–238.

59. Checking Interphone Equipment

It is the duty of each crew member to check his interphone

equipment. He should see that it is complete, in good working order, clean, and properly maintained. Any difficulties should be reported to the tank commander.

60. Use of Definite Terminology

Terminology used by tank commanders in controlling their crews is set forth in paragraph 61. Failure to use standard, specific interphone language causes misunderstanding and disorder. Adherence by all crew members to this standard language is essential to efficient operation of the tank.

61. Interphone Language

a. Terms.

Tank commander	TANK COMMANDER.
Driver	DRIVER.
Gunner	GUNNER.
Loader	LOADER.
Any tank	TANK.
Any unarmored vehicle	TRUCK.
Any antitank gun or artillery piece.	ANTITANK.
Infantry	TROOPS.
Machinegun	MACHINEGUN.
Airplane	PLANE.
Any other target	Briefest descriptive word or phrase.

b. Commands for Movement of Tank.

To move forward	DRIVER MOVE OUT.
To halt	DRIVER STOP.
To reverse	DRIVER REVERSE.
To increase speed	DRIVER SPEED UP.
To decrease speed	DRIVER SLOW DOWN.
To turn right (left)	DRIVER RIGHT (LEFT) STEADY . . . ON.
To turn right (left) 180°	DRIVER RIGHT (LEFT) ABOUT STEADY . . . ON.
To pivot right (left)	DRIVER PIVOT RIGHT (LEFT) STEADY . . . ON.
To move toward a terrain feature or reference point, tank being headed in proper direction.	DRIVER MARCH ON WHITE HOUSE (HILL, DEAD TREE, ETC.).

To follow in column	DRIVER FOLLOW THAT TANK (DRIVER FOLLOW TANK B-9).
To follow road or trail to the right (left).	DRIVER RIGHT (LEFT) ON ROAD (TRAIL).
To start engine	DRIVER TURN IT OVER.
To stop engine	DRIVER CUT ENGINE.
To proceed in a specific transmission range.	DRIVER LOW (HIGH) RANGE.
To proceed at same speed	DRIVER STEADY.

   c. Commands for Control of Turret.

To traverse turret	GUNNER TRAVERSE RIGHT (LEFT).
To stop turret traverse	STEADY . . . ON.

   d. Fire Commands. See chapter 5.

Section IV. CREW DRILL

62. Dismounted Drill

   a. To Form the Tank Crew. Being dismounted, the crew take dismounted posts at the command FALL IN.

   b. Fall In. On command, the crew fall in at attention. The tank commander takes his post two yards in front of the right track, facing to the front. The gunner, driver, and loader, in that order, take posts to the left of the commander at close interval.

   c. To Break Ranks. Being at dismounted posts, the crew break ranks at the command FALL OUT. Crew members habitually fall out to the right of the tank.

   d. To Call Off. Being at dismounted posts, at the command CALL OFF, the members of the crew call off in turn as follows:

Tank Commander	TANK COMMANDER.
Gunner	GUNNER.
Driver	DRIVER.
Loader	LOADER.

   e. To Change Designations and Duties.

      (1) The crew being at dismounted posts, at the command TANK COMMANDER (GUNNER) (DRIVER) FALL OUT:

(a) The crew member designated to fall out moves along the rear of the rank to the left flank position and becomes loader.

(b) The crew members on the left of the vacated post move smartly to the right one position and prepare to call off their new assignments.

(c) The acting tank commander starts calling off as soon as the new crew is re-formed in line.

(2) The movement may be executed by having any member of the crew fall out except the loader.

(3) All movements should be executed at double time with snap and precision.

63. To Mount the Tank Crew

This drill starts with the crew at dismounted posts.

Note. All phrases of crew drill begin with the tank gun forward.

Tank commander	Gunner	Driver	Loader
Command: PREPARE TO MOUNT.			
About face.	About face.	About face.	About face.
Command: MOUNT.			
Stand fast.	Mount right fender.	Stand fast.	Mount left fender.
Mount right fender.	Mount right sponson.	Mount left fender.	Mount left sponson.
Mount right sponson.	Enter turret and take post.	Enter driver's hatch.	Enter turret and take post.
Enter turret and take post.	Connect breakaway plugs.	Turn on master switch.	Turn on radio.
Connect breakaway plugs.		Connect breakaway plugs.	Connect breakaway plugs.
Command: REPORT.			
	Report: GUNNER READY.	Report: DRIVER READY.	Report: LOADER READY.

64. To Close and Open Hatches

a. To Close Hatches. The crew will be at mounted posts.

Tank commander	Gunner	Driver	Loader
Command: 　CLOSE 　HATCHES. Close hatch. Command: 　REPORT.	Insure that turret and tube do not block driver's hatch. Report: 　GUNNER 　READY.	Close hatch. Report: 　DRIVER 　READY.	Close hatch. Report: 　LOADER 　READY.

b. To Open Hatches. The crew will be at mounted posts.

Tank commander	Gunner	Driver	Loader
Command: OPEN 　HATCHES. Open hatch. Command: 　REPORT.	Insure that turret and tube do not block driver's hatch. Report: 　GUNNER 　READY.	Open hatch. Report: 　DRIVER 　READY.	Open hatch. Report: 　LOADER 　READY.

65. To Dismount Tank Crew

This drill starts with the crew at mounted posts, hatches open, and the tank gun forward.

Tank commander	Gunner	Driver	Loader
Command: 　PREPARE TO 　DISMOUNT.			

Tank commander	Gunner	Driver	Loader
Disconnect breakaway plugs. Command: DISMOUNT. Emerge from turret. Move to right sponson. Move to right fender. Take dismounted post and command: CALL OFF.	Disconnect breakaway plugs. Remain in position. Emerge from turret. Move to right sponson. Move to right fender. Take dismounted post.	Disconnect breakaway plugs. Turn off master relay switch. Emerge from hatch. Move to left fender. Take dismounted post.	Disconnect breakaway plugs. Emerge from turret. Move to left sponson. Move to left fender. Take dismounted post.

66. To Dismount Through Escape Hatch

Tank commander	Gunner	Driver	Loader
Command: THROUGH ESCAPE HATCH, PREPARE TO DISMOUNT. Disconnect breakaway plugs. Command: DISMOUNT. Secure carbine and pass it to driver. Move to left side of turret.	Traverse turret to give access to driver's compartment. Disconnect breakaway plugs. Move to left side of turret. Stand fast.	Disconnect breakaway plugs. Turn of master relay switch. Position driver's seat. Release escape hatch. Dismount through escape hatch with carbine. Take dismounted post.	Disconnect breakaway plugs. Turn off radio. Secure submachine gun. Dismount through escape hatch with submachine gun. Take dismounted post.

Tank commander	Gunner	Driver	Loader
Stand fast.	Dismount through escape hatch. Take dismounted post.		
Dismount through escape hatch. Take dismounted post.			

67. Pep Drill

To vary the drill routine and to maintain the interest of the crew members, unexpected periods of pep drill are introduced into the training. Pep drill consists of a series of precision movements executed at high speed and terminating at the position of attention, either mounted or dismounted. For example, the crews being dismounted, the platoon leader commands: IN FRONT OF YOUR TANKS ... FALL IN; MOUNT; DISMOUNT; FALL OUT TANK COMMANDER; ON THE LEFT OF YOUR TANKS ... FALL IN; FORWARD, MARCH; BY THE RIGHT FLANK ... MARCH; TO THE REAR ... MARCH; MOUNT. Preparatory commands for mounting and dismounting are eliminated in this drill. Posts of all crew members are changed frequently.

Section V. SERVICE OF THE PIECE

68. General

a. The gun crew in the tank consists of the loader, who loads the tank gun and the coaxial machinegun; the gunner, who aims and fires the gun; and the tank commander, who controls the fire and, when necessary, adjusts the fire.

b. Teamwork, coordination, and precision of movement are of utmost importance in service of the piece. Crew cooperation in training will provide a smooth, efficient operation in combat when speed is essential and delays or mistakes may be fatal.

69. Gun Crew Positions, Mounted

a. Tank commander _____ Right rear of turret.
b. Gunner _____ Right side of tank gun.
c. Loader _____ Left side of tank gun.

70. Safety Precautions

a. Safety precautions and proper operating procedures are absolutely necessary if the tank is to be kept in operation. The procedures and precautions listed below should be repeated until the normal procedure is a safe procedure.

 b. The loader will—

  (1) Check the breechlock crank stop to insure that it is in the locked position.

  (2) Check the bore of the gun for obstructions prior to and during firing.

  (3) Not allow the fuzed nose or the primer of the round to strike any solid object in the turret.

  (4) Carefully examine each round of ammunition to see that it is clean and not bulged or dented.

  (5) Not attempt to remove ammunition from stowage racks until the loader's traverse safety switch is turned to the OFF position on M41 tanks. On the M41A1 he will use the turret motor switch as a traverse safety.

  (6) Not attempt to disassemble any portion of a round of tank ammunition unless ordered to install concrete-piercing fuze on HE ammunition.

  (7) Stay clear of the path of recoil during and after loading the gun.

  (8) Not attempt to trip the extractors with his fingers when closing the breech.

  (9) Not remove the coaxial machinegun from the tank until it has been cleared and inspected by the tank commander.

  (10) Turn the radio off before the tank engine is started.

 c. The gunner will—

  (1) Always warn the crew before firing the tank gun or coaxial machinegun. During training, he will pause one second after announcing ON THE WAY.

  (2) Alert the crew before traversing the turret in power.

  (3) Release the hand firing lever after firing the gun manually to avoid injury to the loader or damage to the gun as the next round is loaded.

  (4) In event of a misfire, turn off the 76-mm gun switch and announce MISFIRE.

  (5) In the event of a stoppage of the coaxial machinegun, turn off the coaxial machinegun switch and announce STOPPAGE.

 d. The tank commander must know and enforce all necessary safety precautions within his tank.

 e. Any individual who observes a condition which makes firing unsafe will immediately call or signal the command CEASE FIRE.

71. To Open the Breech

Grasp the grip portion of the operating handle, release the latch on the grip, and pull the handle to the rear and down. *When the breech is locked open, immediately return the operating handle to its latched position.*

72. To Load the Gun

a. Open the breech, and return the operating handle to its latched position (check engagement of latch).

b. Select a round of ammunition; grasp it by the base of the shell case with the right hand and by the rear of the ogive with the left hand.

c. Place the projectile in the breech recess, exercising care not to strike the fuze. Move the round forward until the projectile rests in the chamber; remove the left hand and push the round until it is started well into the chamber. Then with the heel of the right hand, fingers closed and joined, vigorously push the round forward into the chamber, rotating the body to the left and sliding the hand off the round above and to the left to insure clearing the breech. The breechblock will automatically push the hand clear when it follows the round into the breech recess. Move to the left side of the turret, clear of the path of recoil. Push in the loader's reset safety (on tanks so equipped), and announce UP.

73. To Unload an Unfired Round or a Misfire

a. To unload an unfired round, the tank commander cups his hands close behind the breech to catch the base of the round as it emerges and to prevent it from dropping to the floor. The loader, by means of the operating handle, opens the breech *slowly*. (*The breech must not be opened rapidly, or the case will become separated from the projectile.*) The loader then removes the round and returns it to its rack.

b. To unload a misfire, the following precautions will be taken: Two more attempts, one electrically and one manually, will be made to fire the piece. Wait 1 minute from the time of the last attempt before opening the breech. Before removing the round, insure that personnel unnecessary to the operation are cleared from the vicinity, then remove the round. Rounds which misfire will not be returned to the racks, but will be moved to a safe place and turned over to ordnance personnel.

74. To Unload a Stuck Round

When a round is stuck in the gun and it is impossible or inadvisable to fire it out, it will be removed. The loader attempts to remove the round with the extracting and ramming tool, placing the tips down and behind the rim of the struck round and applying pressure. If this attempt fails, the round is removed with the bell rammer. With the breech open, the loader takes position to receive the round as it is pushed from the chamber. The tank commander dismounts, inserts the bell rammer into the muzzle of the gun, and pushes it gently through the bore until it is seated on the ogive of the projectile. Exerting a steady pressure, he shoves the round clear so that it may be removed by the loader. To the maximum possible extent, personnel should keep all parts of their bodies clear of the muzzle or breech during the operation. If this procedure fails to remove the round, experienced ordnance personnel should be called. Sometimes the round can be pried out by use of an extracting and ramming tool.

75. To Remove a Stuck Projectile

If the case and projectile become separated, despite care in opening the breech, the chamber will be filled with rags to form a cushion so that the projectile will not damage the breechblock. The breech will be closed, and the procedure described in paragraph 74 will be followed. After the projectile is free in the chamber, the breech is opened and the projectile removed and disposed of in accordance with existing regulations. The chamber must be cleaned.

Section VI. MOUNTED ACTION

76. General

Prior to mounted action drill, the following conditions must be met:

a. Crew mounted.

b. Hatches open.

c. Tank gun forward.

d. Turret-mounted machinegun uncovered.

e. Ammunition stowed.

77. Prepare To Fire

A series of checks of the turret components must be systematically performed by the tank crew to insure that the equipment is in proper working condition. These checks are performed before every operation. During training, the tank crew are drilled to perform these duties to insure coordination of effort, completeness of checks, and speed of execution. All checks listed must be performed in either the assembly area or the attack position. A final check is made just prior to crossing the line of departure. The items marked by asterisks *must* be included in final check. Commands and duties of crewmen are listed below.

Tank commander	Gunner	Driver	Loader
Command: PREPARE TO FIRE. *Clean loader's, gunner's, and tank commander's periscopes and vision blocks. *Clean and inspect M20 periscope (interior).	Check recoil oil by physically feeling indicator tape. *Clean and inspect M20 periscope (interior). Inspect instrument lights and install batteries. *Clean and inspect M97 telescope (interior).	Clean periscopes.	Open breech. Inspect tube and chamber. Close breech. Check and adjust headspace on coaxial machine gun.
Command: *CHECK FIRING SWITCHES.	Turn manual safety to OFF position. Turn 76-mm gun switch to ON position.	Lower seat. Close hatch. Start auxiliary engine.	Push loader's reset safety on tanks so equipped. *Watch action of solenoid and listen for click of percussion mechanism. Recock main gun after each firing check.
*Check firing trigger on power control handle.	*Check firing trigger on manual elevation control handle. *On M41 also check trigger on power control handle. Check manual firing control. Turn off 76-mm gun switch. Turn coaxial machinegun switch to ON position.		Close cover and cock coaxial machinegun. *Watch action of solenoid and listen for strike of firing pin.

Tank commander	Gunner	Driver	Loader
	*Check firing trigger on manual elevation control handle. *On M41 also check trigger on power control handle. Turn off coaxial machinegun switch.		Recock coaxial machinegun after each firing check.
*Check firing trigger on power control handle. Command: *CHECK POWER CONTROL.	On M41A1, check oil in power traverse and elevation systems. *Unlock turret. *Check manual traverse (to insure free movement of turret). *Check manual elevation (on M41A1, adjust tension finger if necessary). *On M41A1, turn turret motor switch and power elevation switch to ON position. *On M41, turn turret motor switch to ON position and dump valve toggle switch to POWER position.		ON M41, check oil in power traversing system. *Check for obstruction to traverse. *On M41, turn loader's traverse safety switch to ON position.
*Check power traverse. On M41A1, check power elevation.	*On M41, check power traverse. (On M41A1, tank commander's control is mechanically linked to gunner's control.		

Tank commander	Gunner	Driver	Loader
Check and adjust headspace and timing on turret-mounted caliber .50 machinegun.	Tank commander's check insures proper functioning of power traverse). Check azimuth indicator for accuracy; traverse turret a complete rotation, stopping at one point to permit loader to check ammunition in hull stowage. Coordinate with crew members; place turret in power and check azimuth indicator for slippage.		Check all ammunition for completeness of stowage and serviceability. Coordinate with gunner in checking ammunition.
Check sight adjustment.	Check elevation quadrant on tanks so equipped. Check sight adjustment. *Set unit battle sight on ballistic unit. Await command to report.		On M41, turn loader's traverse safety switch to OFF position, and on M41A1, turn turret motor switch to OFF position, while checking hull stowage. *Note.* On M41A1, loader may use turret motor switch as traverse safety. *Half-load coaxial machinegun. *Open breech. Await command to report.
*Half-load turret-mounted caliber .50 machinegun.			
Command: REPORT.	Report: GUNNER READY.	Report: DRIVER READY.	Report: LOADER READY.

78. Duties in Firing or Gun Drill

A tank crew must be drilled in the performance of their firing duties to insure coordination of effort and speed of execution. Gun drill is conducted in the form of nonfiring exercises against both stationary and moving targets. Speed must be emphasized throughout this phase of drill. Periods of gun drill must be periodically scheduled and conducted in order to maintain a high standard of tank crew proficiency. To stimulate interest, the tank should move a few yards between each nonfiring exercise, preferably over a simulated combat course in which there are various types of targets that become visible as the tank advances along the course. For the moving target phase of gun drill, a target mounted on a ¼-ton truck can be used. The speed and direction of travel of the target or target vehicle should be varied. The general firing duties of the crew are listed below. For specific firing duties in response to fire commands, see chapter 5.

Tank commander	Gunner	Driver	Loader
Be alert for targets. Control operation of tank by interphone.	Observe in assigned sector.	Observe terrain for best routes. Avoid unnecessary obstacles. Be alert for commands from tank commander.	Observe in assigned sector.
Give fire commands and lay tank gun for direction.			
	Index range on ballistic unit corresponding to announced ammunition. Lay tank gun for deflection and elevation. Fire on target. Adjust fire for target destruction.	Observe in assigned sector.	Load ammunition as announced in fire command. Press loader's reset safety on tanks so equipped. Announce UP.
Observe fire and give subsequent fire commands if necessary.			Continue to load until CEASE FIRE is announced.

Tank commander	Gunner	Driver	Loader
If gunner is unable to see target, adjust fire. Fire turret-mounted caliber .50 machinegun as necessary.	Announce MISFIRE if tank gun fails to fire and STOPPAGE if coaxial machinegun fails to fire.		Follow correct procedure in event of misfire or stoppage. Fire coaxial machinegun manually when directed by gunner. Refill ready racks when necessary. Keep record of ammunition fired.

79. To Clear and Secure Guns

The clear-and-secure-guns procedure, like other procedures in tank operations, is conducted as a drill during training to insure that each crewman knows the duties that he must perform in the clearing of the tank weapons and preparing the tank for an administrative move. If it is desired only to clear the tank weapons, the command is CLEAR GUNS. When the weapons are already cleared and it is desired to secure them, the command is SECURE GUNS. When it is desired to perform both phases together, the command is CLEAR AND SECURE GUNS. The crewmen's duties are listed below:

Tank commander	Gunner	Driver	Loader
Command: CLEAR AND SECURE GUNS. Clear turret-mounted machinegun. Insert T-block.	Turn off firing switches and turret motor switch. Inspect periscope and telescope.	Shut off auxiliary engine.	Clear coaxial machinegun. Insert T-block. Clear tank gun; inspect tube and close breech.

Tank commander	Gunner	Driver	Loader
Place cover on turret-mounted machinegun; secure gun in travel lock.	Coordinate with loader in rearrangement of ammunition stowage.		Fill ready racks.
Assist gunner in placing gun in travel lock.	Place gun in travel lock. Place cover on azimuth indicator.		Secure gun in travel lock.
	Assist loader in placing breech cover on tank gun; turn off instrument lights and remove batteries. Await command to report.	Place muzzle cover on tank gun. Await command to report.	Place breech cover on main gun. Await command to report.
Command: REPORT.			
	Report: GUNNER READY.	Report: DRIVER READY.	Report: LOADER READY.

80. To Load Weapons

The tank weapons are loaded on command. This is normally the fire command. The unit SOP may state the type of ammunition to be carried in the chamber of the tank gun before a target appears. The machineguns are half-loaded at the command PREPARE TO FIRE. In combat, the machineguns will be fully loaded when the unit deploys for action.

81. Stowage and Handling of Ammunition

The ammunition stowage racks (figs. 45 and 46) in the M41 and M41A1 tanks are located in the hull. The ready racks are on the turret floor. The unit SOP should state the number of each type of round to be carried in the ready racks for various combat situations. The rounds carried in the hull should be positioned so the most critical type of ammunition will be readily available to the loader. Ammunition should be handled in a manner which will prevent striking the projectile or primer of the round against a hard surface. Each round must be inspected for dents or bulges and for dirt before it is stowed in the tank. Because the primer is the most sensitive portion of a round of ammunition, the ammunition should be passed into the tank with the primer up.

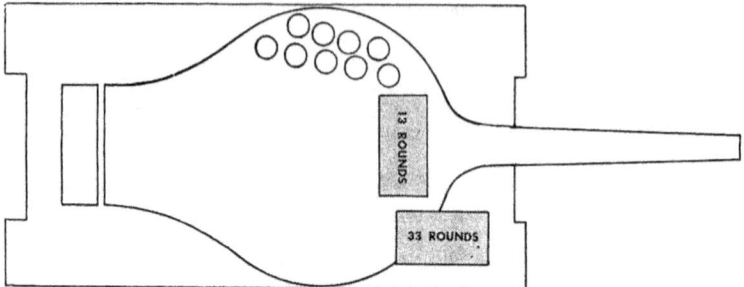

Figure 45. Ammunition stowage racks, M41.

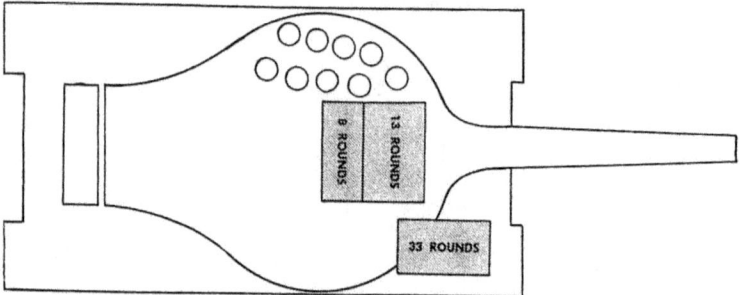

Figure 46. Ammunition stowage racks, M41A1.

Section VII. DISMOUNTED ACTION

82. To Provide Local Security, Dismounted

In certain situations—for example, in assembly areas or under similar static conditions—it may be desirable to provide local dismounted security for the tank. A means of providing dismounted security, without detracting from the ability of the tank to go into action immediately, is for the tank commander to designate a crewman to dismount with hand grenades and carbine. Regardless of which crewman is designated, the remaining crewmen will occupy the positions of the driver, loader, and tank commander.

83. To Abandon Tank

If it becomes necessary to abandon the tank, the crew proceeds as follows:

a. If time permits deliberate action, the tank commander displays the flag signal DISREGARD MY MOVEMENTS (FM 21–60), and supervises the disabling of those weapons remaining in the tank. Back plates are removed from machineguns left in the tank, and the percussion mechanism is removed from the tank gun. Like items in spare parts kits are also removed. Individual weapons and maximum ammunition loads are carried.

b. Ordinarily the tank is abandoned as a result of a direct hit which causes it to catch fire or which disables it so that it becomes a vulnerable target. At the command ABANDON TANK, crew members open the hatches, dismount, and take cover a safe distance from the tank. The tank comander dismounts with the carbine and covers the movement of his crew. The loader dismounts with the submachinegun. In case of fire, it is particularly important to hold the breath until clear of the vehicle; inhaling fumes and smoke may cause serious injury.

84. To Destroy Tank

When the command DESTROY TANK is given, crew members first remove all equipment that is to be carried. They then destroy the tank and the remaining weapons, ammunition, and equipment as prescribed by the unit SOP (see par. 90).

Section VIII. EVACUATION OF WOUNDED FROM TANKS

85. General

Wounded members of the tank crew normally will evacuate themselves from a disabled tank or be removed by their fellow crew members. The utmost speed is necessary in order to save the lives of those who are unhurt as well as the life of the casualty. A burning tank can trap the crew members in a matter of seconds; therefore it is essential that all crewmen become extremely proficient in utilizing the quickest methods of removing each other from the tank. If the action has ceased momentarily, or if the tank has been able to disengage itself without hindering the accomplishment of the mission, the casualty is removed immediately and carried to a protected place, where emergency first aid is administered. Otherwise, the action is continued until an opportunity is presented to remove casualties.

86. Methods of Evacuation

Methods of evacuation described herein are based on the employment of a two-man team, the largest team that can work effectively around a single hatch opening. In some cases, a third man will be able to give considerable help from the inside by placing belts around the wounded man or by moving him to a position where he can be grasped from above. The necessity for swift action usually will require that the casualty be grasped for removal by portions of his clothing or by the arms. If a limb is broken, or if there are other injuries which will be aggravated by these procedures, and if time allows, a lifting sling may be improvised to remove the crewman. Any equipment which is immediately available, such as pistol belts, web belts, or field bag straps, will be used for this purpose.

87. Evacuation Drill, General

a. This paragraph contains general information which may be used as a guide in practicing the evacuation of crew members from any position. During drill, the composition of the evacuating team should be changed frequently to provide practice for all members of the crew in meeting various emergencies.

b. The member of the crew who discovers a wounded crewman announces DRIVER (LOADER, etc.) WOUNDED. If the tank is not actively engaged and the tank commander decides that evacuation is necessary, he commands: EVACUATE DRIVER (LOADER, etc.). Crew members dismount, unless one man is needed to help from inside, and the two nearest the hatch above the wounded man (No. 1 and No. 2 in pars. 88 and 89) take stations at the hatch to act as the evacuation crew. If the man nearest the casualty sees that his help is needed, he stays inside and arranges a sling immediately or takes whatever steps he can to speed the operation. First aid is administered, and the wounded man is moved to a sheltered position. The tank commander reports the casualty.

88. Procedure To Evacuate Casualty From Driving Compartment

Tank commander commands: EVACUATE DRIVER. Loader unlocks driver's hatch from the inside. Tank commander opens the driver's hatch from the outside.

No. 1 (Tank commander)	No. 2 (Loader)
Take position on edge of hatch.	Take position on edge of hatch.
Reach into hatch and grasp hands of casualty, straightening him in seat if necessary.	
Cross casualty's arms over his chest.	Grasp nearest hand of casualty when his arms are crossed.
Raise and rotate casualty so that he faces to the rear.	Raise casualty and help rotate him.

No. 1 (Tank commander)	No. 2 (Loader)
Seat casualty on front rim of hatch; support him in this position while No. 2 jumps to ground.	Help seat casualty; jump to ground.
Lower trunk of casualty into arms of No. 2.	Receive and support trunk of wounded man, holding him beneath arms, around chest.
Lift legs out of hatch as No. 2 lowers casualty along slope plate.	Lower casualty along slope plate and support him until No. 1 can reach ground and assist.
Jump to ground; help No. 2 place casualty in carry position.	Place casualty in carry position.
Carry casualty to protected area.	Help No. 1 carry casualty to protected area.

89. Procedure To Evacuate Casualty From Turret

This procedure can be followed in evacuating any turret crewman. The loader will be used in this example. The tank commander commands: EVACUATE LOADER, and dismounts to the rear deck, where he acts as No. 1. The driver acts as No. 2. The gunner, in this example, assists by positioning straps or belts about the loader so as to obtain the maximum leverage. The gunner unlocks the loader's hatch and opens it with the aid of No. 1.

No. 1 (Tank commander)	No. 2 (Driver)
Take position on turret beside loader's hatch.	Raise casualty as high as possible in hatch opening, holding him around chest.
Grasp casualty under arms.	
Raise casualty through hatch, and seat him on rear edge.	Help No. 1 raise casualty by lifting from below.
Hold casualty while No. 2 dismounts to rear deck.	Dismount to rear deck.
Pick casualty up in arms; carry to rear and lay along back edge of deck.	Help No. 1 pick up casualty and carry to rear of tank; jump to ground.
Help No. 2 lift trunk of casualty off tank; jump to ground.	Lift upper part of casualty's body off tank and support until No. 1 arrives to help.
Lift casualty's hips and legs off tank.	
Carry casualty to protected area.	Help carry casualty to protected area.

Section IX. DESTRUCTION OF EQUIPMENT

90. General

a. The destruction of materiel is a command decision to be carried out only on authority delegated by the division or higher commander. This usually is made a matter of standing operating procedure. *Destruction is ordered only after every possible measure for preservation or salvage of the materiel has been taken and when in the judgment of the military commander concerned, such action is necessary to prevent:*

   (1) Its capture intact by the enemy.
   (2) Its use by the enemy, if captured, against our own or allied troops.
   (3) Its abandonment in the combat zone.
   (4) Knowledge of its existence, functioning, or exact specifications from reaching enemy intelligence agencies.

b. The principles followed are—

   (1) Methods for the destruction of materiel subject to capture or abandonment in the combat zone must be adequate, uniform, and easily followed in the field.
   (2) Destruction is as complete as possible within limitations of time, equipment, and personnel available. If thorough destruction cannot be completed, the most important features of the materiel are destroyed, and parts which cannot be easily duplicated and are essential to the operation or use of the materiel are ruined or destroyed. *The same essential parts are destroyed on all like units to prevent the enemy from constructing a complete unit from several damaged ones.*

c. Crews are trained in employing prescribed methods of destruction. *Training does not involve actual destruction of materiel.*

d. Certain methods of destruction require special tools and equipment, such as TNT and incendiary grenades, which may not be items of issue. The issue of such special tools and equipment and the conditions under which destruction will be effected are command decisions, and depend upon the tactical situation.

e. The proper methods for the destruction of the OVM and the tank are covered in TM 9-730.

CHAPTER 4
CREW PREVENTIVE MAINTENANCE

Section I. INTRODUCTION

91. General

The tank commander is responsible for insuring that required preventive maintenance is performed. Mechanical efficiency is essential to the operation of tank units; therefore, each tank must be serviced systematically at intervals during each day of use. Defects can then be discovered and corrected before they result in mechanical damage or failure. Crew members make their individual inspections and report the results to the tank commander, who, in his own report, lists all items requiring the services of maintenance personnel. In supervising crew maintenance or other services performed at periodic intervals and from day to day, the tank commander delegates responsibility to crew members as necessary. Maintenance procedures omitted from this manual are set forth in TM 9-730.

92. Maintenance To Be Performed

a. Inspections are made of all personal equipment and weapons, fire-control equipment, communication equipment, vehicle equipment, and mechanical features of the vehicle. Inspections of instruments, radios, lights, tracks, suspension system, and engine performance are made in accordance with provisions of appropriate technical manuals. The driver fills in DD Form 110 (Vehicle and Equipment Operational Record), indicating thereon deficiencies and maintenance work required, and the loader fills in DA Form 11-238 (Operator First Echelon Maintenance Check List). Any irregularities entered on these forms which are not repaired before the tank is again used will be re-entered continually until the deficiency has been corrected.

b. In succeeding paragraphs, the duties of the crew members are given in chart form to facilitate preliminary training of the soldier in performing crew maintenance. It is not intended that these procedures be followed as precision exercises, such as in crew drill or gun drill; they are a guide for performing crew maintenance in a logical and efficient manner with a minimum loss of time.

Section II. CREW MAINTENANCE PROCEDURES

93. Before-Operation Service

This inspection begins with the tank locked and covered by a tarpaulin, with the gun in the traveling position. The turret is traversed as necessary to facilitate the various operations. A minimum of 2 man-hours is required to properly perform this service.

Tank commander	Gunner	Driver	Loader
Command: FALL IN; PREPARE FOR INSPECTION.	Form in front of tank.	Form in front of tank.	Form in front of tank.
Inspect crew. Command: PERFORM BEFORE-OPERATION SERVICE. Supervise inspection and filling out of DD Form 110 and DA Form 11-238.	Stand inspection. Untie tarpaulin ropes.	Stand inspection. Fill out DD Form 110 during inspection.	Stand inspection. Fill out DA Form 11-238 during inspection.
Remove, fold, and stow tarpaulin.	Help tank commander.	Inspect ground beneath tank for fuel and oil leaks. Inspect tracks and suspension.	Help tank commander.
Supervise.	Mount right front fender. Open loader's hatch and enter tank. Unlock tank commander's and driver's hatches.	Procure hand tools and lay out on left front fender.	Mount left front fender. Check contents of stowage boxes, pioneer tools, and tow cable.
Supervise.	Remove breech cover; inspect breech of 76-mm gun and replenisher indicator. Pass all covers to the loader. Unlock turret lock.	Mount tank via left front fender and check fuel.	Remove and stow muzzle cover. Unlock gun travel lock. Check stowage and lashing of personal gear.

Tank commander	Gunner	Driver	Loader
	Check main and auxiliary engine air filter and oil level in hydraulic reservoir. Dismount.	Open engine compartment grills; check oil level in main and auxiliary engines. Check transmission for presence of oil.	Receive and stow covers (breech and azimuth); mount antenna. Help driver.
Procure cleaning rods and material. Check lights.	Help swab bore of 76-mm gun and MGs. Take mounted post. Inspect sights and turret controls. Traverse turret to check azimuth indicator.	Help swab bore of 76-mm gun and MGs. Take mounted post. Check fixed fire extinguisher; close master switch; start auxiliary engine; check operation of heater, steering controls, lights, horn, bilge pumps, drain valves, and fuel cutoff.	Help swab bore of 76-mm gun and MGs. Take mounted post. Check portable fire extinguisher; operation of ventilator blower; water; first-aid kit; rations, 76-mm armament tools, spare parts, spare hydraulic recoil, and engine oil.
Observe main engine exhaust; make visual check for oil leaks, vibration and unusual noises.	Help attach shell bag; check cleanliness of chamber and breech of tank gun.	Start main engine, stop auxiliary engine. During warmup period check instruments, smoothness of operation, any unusual noises.	Attach empty cartridge bag; mount coaxial MG and adjust headspace on MG; check mount and firing mechanism.
Check external interphone.		Turn on external interphone. After warmup period, check magneto operation.	Turn on radio; check with tank commander on external interphone.

Tank commander	Gunner	Driver	Loader
Check transmission oil level; close grills over engine compartment.	Check headspace on Cal .50 MG.	Check all publications pertinent to the vehicle (TM, LO, Standard Form 91, DD Form 518).	Check hand grenades and Cal .45 ammo (if applicable).
Direct driver to move tank forward two tank lengths. Inspect right, then left, track and suspension. Direct driver to move to original position. Check tools and stow.	Clean periscopes and vision blocks.	Take command from tank commander.	Finish operator daily maintenance on radio equipment as prescribed in DA Form 11-238.
Take mounted post. Check carbine and ammunition, power control, binoculars, flag set, operation of radio.	Traverse turret to traveling position or firing position.	Check submachinegun and ammunition, periscopes, and escape hatch.	Lock gun travel lock, if gun is put in travel position.
Connect breakaway plugs. Check interphone system by commanding: REPORT. Report READY to platoon leader.	Connect breakaway plugs. Report: GUNNER READY.	Connect breakaway plugs. Report: DRIVER READY.	Connect breakaway plugs. Report: LOADER READY.

94. During-Operation Checks

This is a continuous process for all crew members during operation of the vehicle. All crew members must remain on the alert at all times for unusual noises and conditions, reporting to the tank commander if any are discovered.

Tank commander	Gunner	Driver	Loader
Check operation of radio and interphone system; observe security of antenna; check turret-mounted Cal .50 machinegun and other visible outside equipment.	Check operation of sighting and fire-control equipment; check elevating and traversing mechanism if gun is in firing position.	Observe instrument panel; check operation of controls.	Check security of equipment in turret, including coaxial machinegun, radio, portable fire extinguisher.

95. At-the-Halt Service

The length of time for the halt is the basis for determining how much of the following service will be completed; priority should be given to items accordingly. The tank commander will be informed of the time allotted for the halt and will indicate to his crew just how much of this time will be allotted to maintenance and inspection and how much for relief of crew members. As the result of training and experience, the crew will learn just about what can be accomplished in a given length of time.

Tank commander	Gunner	Driver	Loader
Command: PERFOM AT-HALT SERVICE. (Use interphone system when applicable.) Disconnect breakaway plugs, supervise inspection.	Disconnect breakaway plugs. Release turret lock; check operation of manual traverse, power traverse, and manual and power elevation.	Disconnect breakaway plugs. Fill out DD Form 110 (during-operation and at-halt sections); idle main engine properly; check instruments.	Monitor radio. Man turret-mounted machinegun (if applicable). If not applicable, check items listed under *loader* in before-operation section. Assist tank commander.

Tank commander	Gunner	Driver	Loader
Dismount to rear deck; unlock gun traveling lock (if applicable); open engine compartment doors, check transmission oil while engine is running; inspect operation of main engine and security of components.	Check sight adjustment, sighting and fire-control equipment, coaxial machinegun and mount, firing mechanisms, security of air cleaners.		
Clean outside of all turret periscopes and vision blocks.		Stop main engine.	
Inspect operation of auxiliary engine; check main engine oil level.		Start auxiliary engine.	
Close engine compartment doors. Lock gun in traveling lock (if applicable); check pioneer tools and tow cable.	Traverse turret to enable commander to lock gun traveling lock.	Stop auxiliary engine. Check driver's compartment for fuel and oil leaks; check fixed fire extinguishers, brakes, and controls.	Check tracks and suspension system.
Dismount; check underneath vehicle for fuel or oil leaks; help driver check lights. Take mounted post.	Check security of equipment stowed in turret.	Check service and blackout lights; clean periscopes.	
	Take mounted post.		Take mounted post when commander mounts.
Command: REPORT.	Report: GUNNER READY.	Report: DRIVER READY.	Report: LOADER READY.

100   AGO 4054B

Tank commander	Gunner	Driver	Loader
Command: FALL OUT FOR BREAK (if applicable, alternate in manning of turret-mounted machinegun). Command: MOUNT UP (if dismounted). Connect breakaway plugs. Command: REPORT.	Take mounted post; connect breakaway plugs. Report: GUNNER READY.	Take mounted post; connect breakaway plugs. Report: DRIVER READY.	Take mounted post; connect breakaway plugs. Report: LOADER READY.

96. After-Operation Service

Immediately after operation, the tank is given the service and maintenance necessary to prepare it in every way for sustained operation. This covers all points listed in the before-operation service and covers them in practically the same order. Obviously, more extensive servicing and maintenance are required. During this operation, the vehicle is cleaned, serviced, and replenished with fuel, oil, grease, ammunition, first-aid equipment, water, and rations. Refer to the vehicle lubrication order for the proper types and amount of oil and greases and intervals of use. All safety precautions against fire must be observed while refueling. A portable fire extinguisher must be available on the rear deck of the tank and must be manned by a crew member. Safety precautions must be carefully observed in handling ammunition.

Tank commander	Gunner	Driver	Loader
Command: PERFORM AFTER-OPERATION SERVICE. Supervise operation.		Fill out appropriate section of DD Form 110 during inspection.	

Tank commander	Gunner	Driver	Loader
	Clean inside turret, tank gun, turret-mounted machinegun, and coaxial machinegun; help replenish ammunition, water rations, and first-aid equipment.	Idle engine properly before stopping; clean outside of vehicle; clean engine compartment and driver's compartment; replenish fuel, oil, and greases.	Help gunner clean inside turret, tank gun, and turret-mounted and coaxial machineguns; replenish ammunition, water, rations, and first-aid equipment.
Check DD Form 110 and DA Form 11–238 for completeness; make final inspection of vehicle; report READY to platoon leader.		Complete DD Form 110 and give to tank commander.	Complete DA Form 11–238 and give to tank commander.

97. Weekly Preventive Maintenance Service

This maintenance service is performed weekly in addition to the daily maintenance services. It is also performed after each field operation in combat and on maneuvers. In garrison operation, allowances should be made for these services in preparing training schedules and work details. In combat and maneuvers, provisions must be made to allow time for crew members to perform this preventive maintenance. A minimum of 16 man-hours is required. The current vehicle technical manual and lubrication order (LO 9–7016) must be followed. The following procedure is a suggested sequence to facilitate preliminary training of the crew.

Tank commander	Gunner	Driver	Loader
Command: FALL IN, PREPARE FOR INSPECTION.	Form in front of tank.	Form in front of tank.	Form in front of tank.
Inspect crew.	Stand inspection.	Stand inspection.	Stand inspection.

Tank commander	Gunner	Driver	Loader
Supervise inspection.	Clean and paint any rusty spots in turret.	Clean engine and engine compartment; make detailed inspection of main and auxiliary engines; service batteries.	Perform radio operator's weekly maintenance as prescribed on DA Form 11-238.
Assist driver; clean and check hand and pioneer tools; clean and spot-paint rusty spots on outside of vehicle.	Tighten all track pads; inspect tracks and suspension system; tighten track pin nuts.	Start main engine and drive vehicle forward as required to tighten track pad nuts; clean and paint rusty spots in driver's compartment.	Help gunner tighten track pad nuts; inspect tracks and suspension system.
Check DD Form 110 and DA Form 11-238 for completeness; make final inspection of vehicle; report READY to platoon leader.	Help lubricate.	Lubricate as required; complete DD Form 110 and give to tank commander.	Help lubricate.

CHAPTER 5
CONDUCT OF FIRE

Section I. INTRODUCTION

98. General

The material contained in this chapter concerns the conduct of fire (direct fire) for the Tank, 76-mm Gun, M41 or M41A1. The proper and timely utilization of the excellent fire-control equipment will enable the tank gun crew to obtain, with exceptional speed, a very high percentage of first-round hits. In order to realize the full capabilities of the tank, the crew must be thoroughly drilled in their firing duties. This training will be conducted prior to any subcaliber or service firing and consists of nonfiring exercises for both stationary and moving targets. After qualification, the drill is conducted at frequent intervals to insure that a high level of crew proficiency is maintained. For principles and techniques on conduct of fire, see FM 17–12.

Section II. FIRING DUTIES

99. General

The effectiveness of the fire of tank weapons is entirely dependent on the coordinated action of the tank crew. Listed below are the general firing duties performed by the tank crew members.

Crew member	Firing duties
Driver	Observes in assigned sector and moves the tank as directed by the tank commander.
Loader	Loads the tank gun and coaxial machinegun; reduces stoppages; inspects, cleans, and stows ammunition; assists in removal of misfires, stuck rounds, and separated rounds; observes in assigned sector.
Gunner	Aims and fires the tank gun and coaxial machinegun, utilizing the M20 periscope or M97 telescope, ballistic unit, gun switches, traversing and elevating controls, and firing trigger. Controls manual safety, adjusts fire of tank gun and coaxial machinegun. Observes in assigned sector.

Crew member	Firing duties
Tank commander	Controls movement of the tank and actions of the crew; observes terrain and selects targets; gives initial fire commands and subsequent fire commands when necessary; lays the gun initially for direction, using the tank commander's power control handle and M20 periscope; supervises and assists gunner in adjusting fire; controls volume of fire; fires caliber .50 turret-mounted machinegun.

Section III. FIRING AT STATIONARY TARGETS

100. Initial Fire Commands and Firing Duties

The tank commander controls the fire of his tank and coordinates the action of his crew by the timely issuance of fire commands. The initial fire command contains the necessary information for the crew to load, aim, and fire the tank weapons. Listed below are examples of initial fire commands, together with the specific duties performed by crewmen in response to the commands under various conditions that might be encountered when firing at stationary targets.

a. *Condition 1.*

(1) Primary fire-control equipment, consisting of M20 periscope and ballistic unit.

(2) Target: stationary tank.

Element	Command	Crewmen's firing duties		
		Tank commander	Gunner	Loader
Alert.	GUNNER.	Using the M20 periscope and power control handle, lays gun for direction and estimates range to the target while announcing the initial fire command.	Turns on turret motor and elevation switches if turret is not in power.	Stands by.
Ammunition.	SHOT.		Turns 76-mm gun switch to ON position and mentally	Selects and loads a round of shot, moves clear of path of recoil, and

Element	Command	Crewmen's firing duties		
		Tank commander	Gunner	Loader
Range.	1100.		notes type of ammunition announced. Indexes the announced range on the proper ammunition scale of the range drum of the ballistic unit.	announces UP. Selects another round of shot.
Direction. Description.	Omitted.* TANK.		Announces IDENTIFIED when he sees the target, and takes control of turret.	
Command to fire.	FIRE.		Makes final precise lay, announces ON THE WAY, and fires.	Loads round of shot and continues to load shot without command until CEASE FIRE or a change in ammunition is announced.

*If power control is inoperative, a direction element must be announced.

b. *Condition 2.*
  (1) Secondary fire-control equipment, consisting of the M97 telescope and the ballistic unit.
  (2) Target: stationary heavy tank.

Element	Command	Crewmen's firing duties		
		Tank commander	Gunner	Loader
Alert.	GUNNER.	Using the M20 periscope and power control handle, lays gun for direc-	Turns on turret motor and elevation switches if not in power.	Stands by.

106   AGO 4054B

| | | Crewmen's firing duties | | |
Element	Command	Tank commander	Gunner	Loader
Ammunition.	HYPER-SHOT.	tion and estimates range to target while announcing the initial fire command.	Turns 76-mm gun switch to ON position and mentally notes type of ammunition announced.	Selects and loads a round of hyper-shot, moves clear of the path of recoil, and announces UP. Selects another round of hyper-shot.
Range.	1,300.		Indexes the announced range* on the HYPER-SHOT scale of the range drum, then reads the range indicated under the index line on the shot scale of the range drum. In this case 900 yards would be indicated.	
Direction. Description.	Omitted.** TANK.		Announces IDENTIFIED when he sees the target, and takes control of turret.	
Command to fire.	FIRE.		Using a range of 900 yards on the telescope, makes	Loads round of hyper-shot and continues to load hyper-

Element	Command	Crewmen's firing duties		
		Tank commander	Gunner	Loader
			final precise lay, announces ON THE WAY, and fires.	shot without command until CEASE FIRE or a change of ammunition is announced.

* When firing any type of ammunition other than that for which the sight is graduated, the gunner must use the ballistic unit, an aiming data chart, or a firing table to determine what range he must apply to the telescope reticle in order to hit the target.

** If power control is inoperative, a direction element must be announced.

101. Sensings

Rounds are sensed in relation to the target. The tank comander and gunner will mentally sense each round for range and deflection. These sensings are not announced unless the gunner fails to observe the burst or tracer through his direct-fire sight, in which case the gunner will announce LOST. The tank commander will then announce his sensing for range only, and his subsequent fire command. The five possible range sensings are—

a. Target. A round is sensed as TARGET only when the round is observed to actually strike the target, causing the target to change shape, pieces to fly off, or the target to completely disappear. When shot strikes a metal object, there is usually a distinctive orange flash.

b. Over. A round is sensed as OVER when the burst appears beyond the target or the tracer passes above the target. This sensing is readily identified when firing HE, since the burst tends to silhouette the target.

c. Short. A round is sensed as SHORT when either the burst or the point of strike is observed between the gun and the target. The target is sometimes temporarily obscured by smoke and/or dust.

d. Doubtful. A round is sensed as DOUBTFUL when the burst appears to be correct for range but off in deflection, or when the tracer passes to the right or left of the target but apparently is correct for range. A range change is not made on a DOUBTFUL sensing; a deflection correction is normally sufficient to secure a target hit.

e. Lost. A round is sensed as LOST when the tank commander or gunner fails to observe the point of strike, burst, or tracer.

The point of strike may not be visible due to obscuration, terrain, or failure of the round to detonate. (Based on the tank commander's terrain appreciation, he may be justified in making a range change.)

102. Adjustment of Direct Fire, Primary Method

The primary method of adjustment is burst-on-target, in which the gunner, observing through his direct-fire sight, notes the point on the sight reticle where the burst or tracer appears in relation to the target and, without command from the tank commander, moves that point of the gun-laying reticle onto the center of the target before firing the next round. This method of adjustment provides a quick, accurate means of obtaining second-round target hits. The gunner uses this method whenever possible. Typical examples of burst-on-target adjustment, using the M20 periscope reticle as well as the telescope reticle, appear below.

a. Situation 1. Primary fire-control equipment, stationary target, HE ammunition.
 (1) The gunner immediately re-laid after firing the first round so that the burst appeared on his sight reticle in its proper relation to the target. This round was off in deflection to the right (fig. 47).
 (2) The gunner mentally noted the point on the sight reticle where the burst appeared and, with the turret controls, moved that point to the center of the target (fig. 48). Without command, he fired and obtained a target hit.

b. Situation 2. Secondary fire-control equipment, stationary target, HE ammunition.
 (1) The gunner noted that the first round fired (fig. 49) struck short of the target and on line with the center of the target.
 (2) The gunner mentally noted the point on the sight reticle where the burst appeared and, with the turret controls, moved that point onto the center of the target (fig. 50). Without command, he fired the next round and obtained a target hit.

103. Adjustment of Fire, Alternate Method

The alternate method of adjustment is the tank commander's means of adjustment when the primary method cannot be effectively used because of obscuration, terrain, or extreme range. The alternate method of adjustment involves the use of standard range changes to be announced by the tank commander under certain conditions. The conditions and standard range changes are as follows:

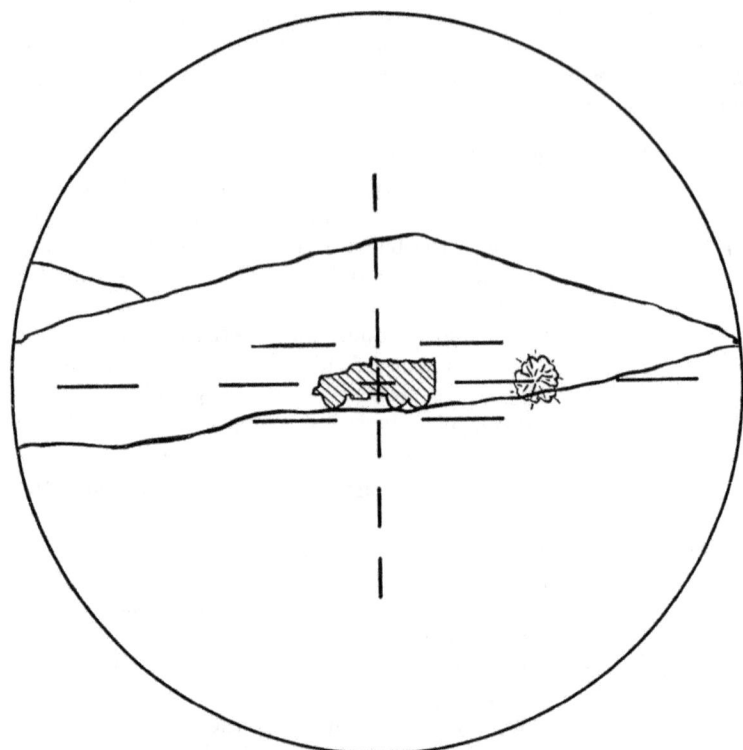

Figure 47. Situation 1—stationary target.

    a. When the estimated range to the target is *1,500 yards or less* and the gunner fails to observe the tracer or burst of his initial round, he will announce LOST. The tank commander will then announce a sensing and subsequent fire command, adding or dropping 200 yards if there was a range error. If the gunner observes this round in his sight, he will apply burst-on-target. However, if this second round is also LOST to the gunner, the tank commander will continue with the adjustment, making whatever deflection and range changes he feels are necessary to obtain a target hit. Deflection errors are measured with the binocular, and range changes are made in multiples of 50 yards. If the necessary range change is less than 50 yards, the command may be ADD (DROP) A HAIR.

    Note. For practical purposes, a 1-mil elevation change will change the range 100 yards. During the alternate method of adjustment, the gunner will use the range lines of the gun-laying reticle to make the necessary range change.

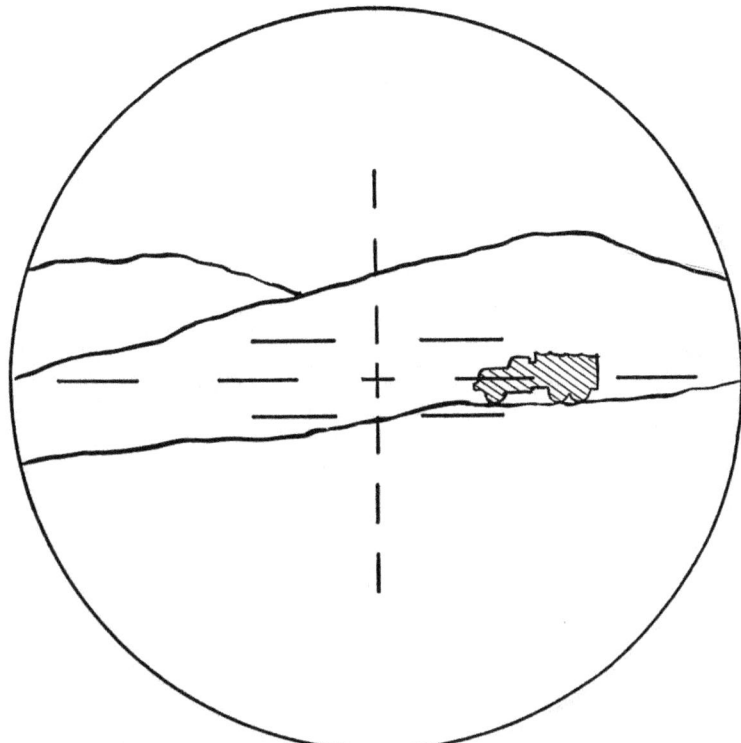

Figure 48. Situation 1—stationary target.

b. When the estimated range to the target is *more than 1,500 yards* and the gunner fails to see the tracer or burst of his initial round, he will announce LOST. The tank commander will then announce a sensing and subsequent fire command, adding or dropping 400 yards for the first adjustment if there was a range error. If the gunner observes this round, he will apply burst-on-target; if not, he will announce LOST, and the tank commander will continue with the adjustment as described in *a* above.

Note. If an extremely large error is made in the initial *estimated* range, the tank commander may cease fire and announce a new initial fire command.

c. When the gunner, during an adjustment, fails to observe a round *after* applying the burst-on-target method, he will announce LOST. The tank commander will take over and use the alternate method of adjustment, making the deflection and range changes he feels are necessary to obtain a target hit.

d. When the gunner and tank commander both fail to observe the round, the gunner will announce LOST. The tank commander

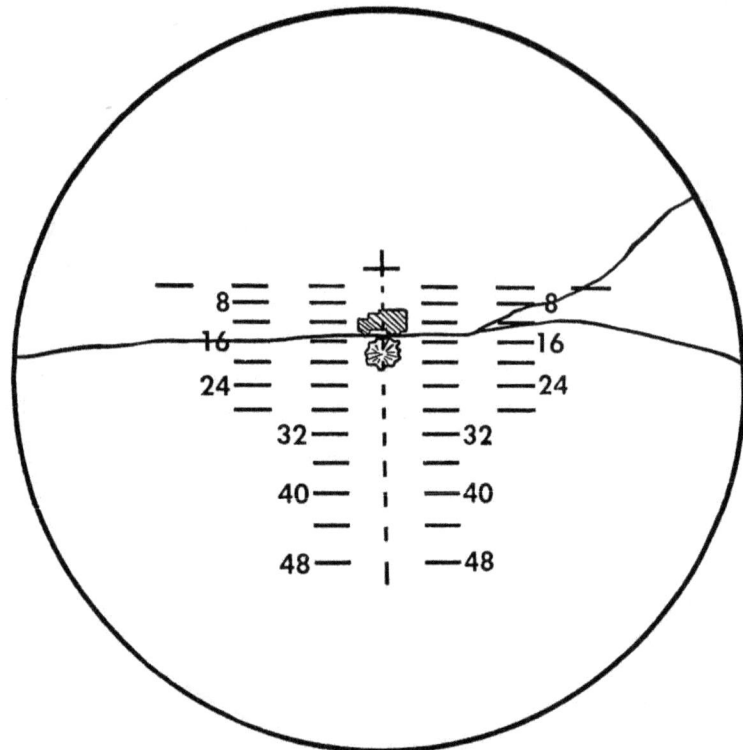

Figure 49. Situation 2—stationary target.

will also announce LOST and will give a subsequent fire command. He may fire another round without changing the range, or he may give a range change to bring the next round to where it can be observed.

e. In all of the above cases, the gunner will apply the announced range change by use of his direct-fire sight and will use the primary method of adjustment whenever possible.

f. The tank commander maintains control of his tank at all times and may take over adjustment of fire at any time. Once an adjustment of the initial round has been made (either by burst-on-target or subsequent command), or a target hit has been obtained, the standard range change no longer applies.

104. Subsequent Fire Commands

a. General. Subsequent fire commands are issued by the tank commander to the tank gun crew to meet various conditions en-

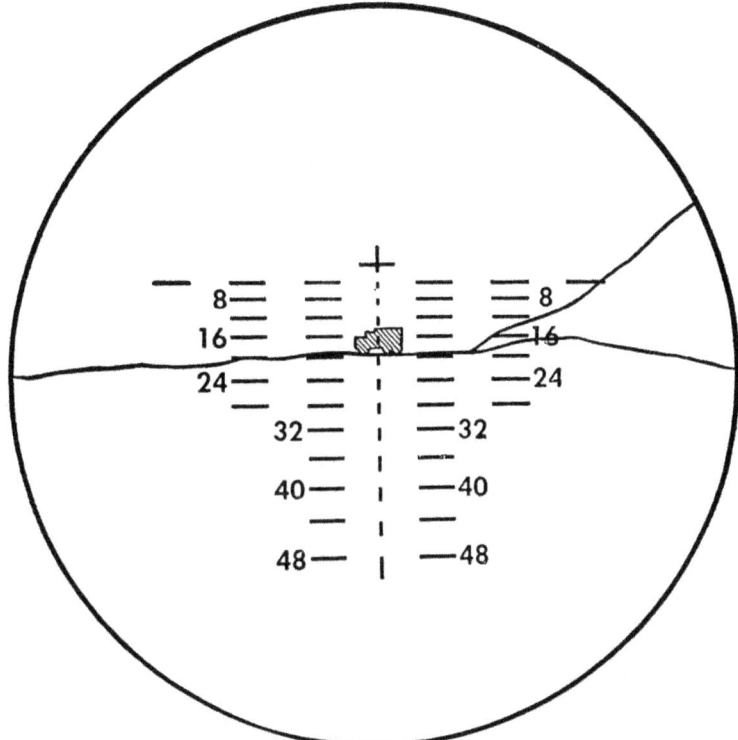

Figure 50. Situation 2—stationary target.

countered during firing. These commands are necessary when the gunner has failed to observe the fired round and has announced his sensing as LOST, when it is desired to change the ammunition or fuze, when the tank commander desires to cease firing, or when the tank commander desires to take over the adjustment of fire for any reason. The sequence of elements of the subsequent fire commands used when firing at stationary targets is as follows:

Elements	*Example*
(1) Deflection change (in mils).	RIGHT 3
(2) Range change (in yards).	ADD 200
(3) Command to fire.	FIRE.

Note. Elements may be omitted from the subsequent fire command if not applicable to an adjustment; therefore, the subsequent fire command may contain one, two, or three elements. If a change of ammunition is desired, it is combined with the command-to-fire element; see *b* below.

b. To Change Ammunition or Fuze. During firing, it may be necessary to designate a different type of ammunition. For example, if a round of shot has penetrated a pillbox or heavy masonry building and the tank commander desires to fire HE through the opening, he commands FIRE HE. This alerts all crewmen to a change in the ammunition. The loader at once loads the HE round, announcing HE-UP to inform the gunner and tank commander that the change has been made. He continues to load HE until he hears CEASE FIRE or another change. The gunner, on hearing the change in ammunition, indexes the correct (new) ammunition on the ballistic unit. The commander uses the same procedure to change the fuze setting; for example, to change from fuze superquick to fuze delay, he commands FIRE FUZE DELAY. Normally, a chambered round will be fired even though a change in ammunition or fuze has been designated.

Section IV. FIRING AT MOVING TARGETS

105. General

When firing at moving targets, the tank gun crew have relatively the same duties as when engaging stationary targets. A moving target is one which has apparent speed. Targets moving across the line of sight, either horizontally or diagonally, have apparent speed. Targets moving directly toward or directly away from the tank have no apparent speed and are not treated as moving targets. Proper leading and tracking are of utmost importance when firing at moving targets.

106. Leading

If the gunner fires a round while the gun is aimed directly at a moving target, the target will move out of the path of the projectile, causing the projectile to miss the target. To compensate for this movement of the target, the gun must be aimed ahead of the target so that the target and projectile will meet. This technique is called *leading*. To aim ahead of the target, the gunner will use the lead lines in his direct-fire sights. A lead equals 5 mils regardless of the range to or speed of the target and is measured from the center of the target. Lead lines in the sight reticle are 5 mils in length.

107. Tracking

In order to maintain the proper lead, the gunner must cause the movement of the gun to keep pace with the movement of the target. This manipulation is called *tracking* and is a combination of traversing and changing elevation in order to maintain proper sight alinement. While the tank commander announces the initial fire command, he lays the gun for direction, using his power control handle and periscope, and continues to track the target. When the gunner sees the target, he announces IDENTIFIED, takes control of the turret, tracks the target with the prescribed lead, announces ON THE WAY, and fires. In tracking, the gun must be traversed through and ahead of the target center until the proper lead is applied. The gunner tracks with a smooth, continuous motion, maintaining a constant sight picture before, during, and after firing. In order that proper sensings and/or adjustments can be made, he will not stop the movement of the gun while he fires; nor will he attempt to ambush the target by moving ahead of it, stopping, and firing when the target reaches the proper lead on the sight reticle.

108. Initial Fire Commands and Firing Duties, Moving Targets

Initial fire commands are the same as those used for engaging stationary targets with the primary fire-control equipment, with the addition that a lead element will be announced just before the command to fire. Normally, one lead will be used initially, regardless of target speed or range. Listed below are examples of initial fire commands, together with the specific duties performed by crewmen in response to each element.

a. Condition 1.
  (1) Primary fire-control equipment, consisting of M20 periscope and ballistic unit.
  (2) Target: moving tank.

Element	Command	Crewmen's firing duties		
		Tank commander	Gunner	Loader
Alert.	GUNNER.	Using his periscope and power control handle, lays gun for direction while estimating the range and announcing the initial fire command. Example: GUNNER SHOT, 1,100, TANK, ONE LEAD, FIRE.	Turns on turret motor and elevation switches if not in power.	Stands by.

Element	Command	Crewmen's firing duties		
		Tank commander	Gunner	Loader
Ammunition.	SHOT.		Makes mental note of type ammunition announced.	Selects and loads a round of shot, moves clear of the path of recoil, and announces UP. Selects another round of shot.
Range.	1,100.		Indexes 1,100 on the shot scale of the ballistic unit.	
Direction.	Omitted.			
Target description.	TANK.		Announces IDENTIFIED as soon as he sees the target, takes control of the turret, and tracks the target.	
Lead.	ONE LEAD.		Takes proper sight picture, applying one lead.	
Command to fire.	FIRE.		Announces ON THE WAY, and fires.	
				Loads round of shot and continues to load shot without command until CEASE FIRE or a change in ammunition is announced.

  b. *Condition 2.*
   (1) Secondary fire-control equipment, consisting of M97 telescope and ballistic unit.
   (2) Target: moving truck.

Element	Command	Crewmen's firing duties		
		Tank commander	Gunner	Loader
Alert.	GUNNER.	Using his periscope and power control handle, lays gun for direction while estimating range and announcing initial fire command. Example—GUNNER, HE, 1,700, TRUCK, ONE LEAD, FIRE.	Turns on turret motor and elevation switches if not in power.	Stands by.
Ammunition.	HE.	Makes mental note of type ammunition announced.		Selects and loads a round of HE, moves clear of the path of recoil, and announces UP. Selects another round of HE.
Range.	1,700.		Indexes 1,700 on the HE scale of the ballistic unit. Reads the range that is indicated by the index line for shot and applies this range, using the telescope reticle.	
Direction. Target description.	Omitted. TRUCK.		Announces IDENTIFIED as soon as he sees the target, takes control of the turret, and	

Element	Command	Crewmen's firing duties		
		Tank commander	Gunner	Loader
Lead.	ONE LEAD.		tracks the target. Takes proper sight picture applying one lead.	
Command to fire.	FIRE.		Announces ON THE WAY and fires.	
				Loads round of HE and continues to load HE without command until CEASE FIRE or a change in ammunition is announced.

109. Sensings, Moving Target

Rounds fired at moving targets are sensed in relation to the target as when firing at stationary targets (par. 101). Any deflection errors are lead errors, and the actual mil error must be converted to leads and/or fraction of leads.

110. Adjustment of Fire, Moving Target

a. Adjustment of Fire, Primary Method. The burst-on-target method of adjustment for moving targets is the same as for stationary targets, and will be used when the gunner is able to sense the rounds fired. Typical examples of burst-on-target adjustment, using the M20 periscope reticle as well as the telescope reticle, appear below.

    (1) *Situation 1.* Primary fire-control equipment, moving target, shot ammunition.

        (*a*) The gunner immediately re-laid after firing the first round and observed the tracer to pass above and to the rear of the target (fig. 51).

        (*b*) The gunner mentally noted the point on the sight reticle where the tracer appeared and, with the turret controls, moved that point to the center of the target

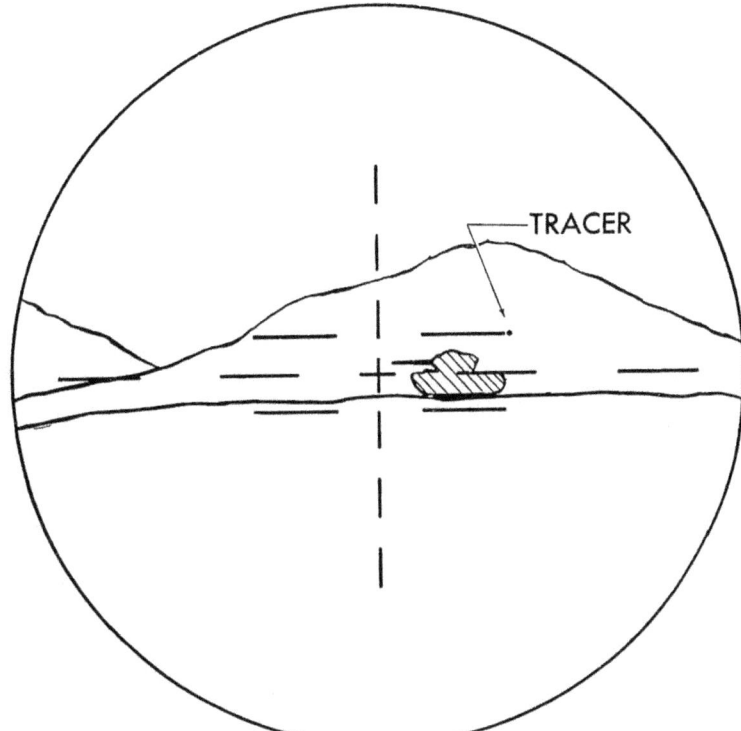

Figure 51. Situation 1—moving target.

(fig. 52). Without further command, he fired and obtained a target hit. Note that he continued to track and increased the lead to compensate for the deflection error.

(2) *Situation 2.* Secondary fire-control equipment, moving target, shot ammunition.

(a) The gunner immediately re-laid and noted that the first round fired was correct for lead but over in range (fig. 53).

(b) The gunner mentally noted that point on his reticle where the tracer appeared to pass over the target. With his turret controls, he placed that point of the reticle on the center of the target (fig. 54), fired, and obtained a target hit.

b. Adjustment of Fire, Alternate Method. The alternate method of adjustment is the tank commander's means of adjusting tank

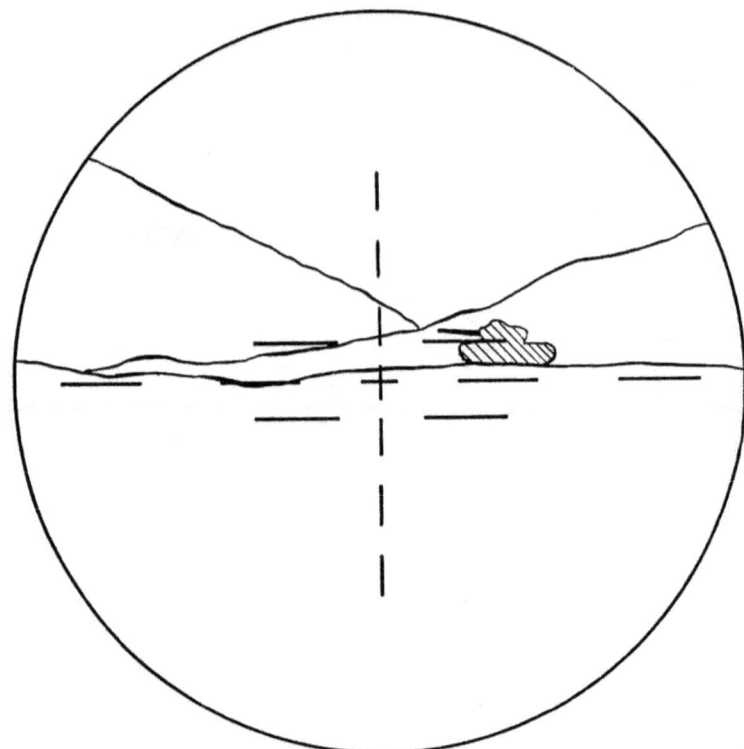

Figure 52. Situation 1—moving target.

fire when the gunner cannot effectively apply the primary method. The conditions under which this method is applied are the same as those for stationary targets, and a subsequent fire command is given.

111. Subsequent Fire Commands, Moving Target

When necessary, the tank commander will issue subsequent fire commands. Range corrections are announced in the same manner as is prescribed for adjusting on stationary targets (par. 104). Lead corrections will be announced as a change in leads rather in mils. For example, if a round passes behind the center of the target, the tank commander may announce ONE MORE or ONE HALF MORE, and the gunner will increase the lead accordingly. Conversely, if the round passes in front of the center of the target, the tank commander may announce ONE LESS or ONE HALF LESS, and the gunner will decrease the lead. The sequence of

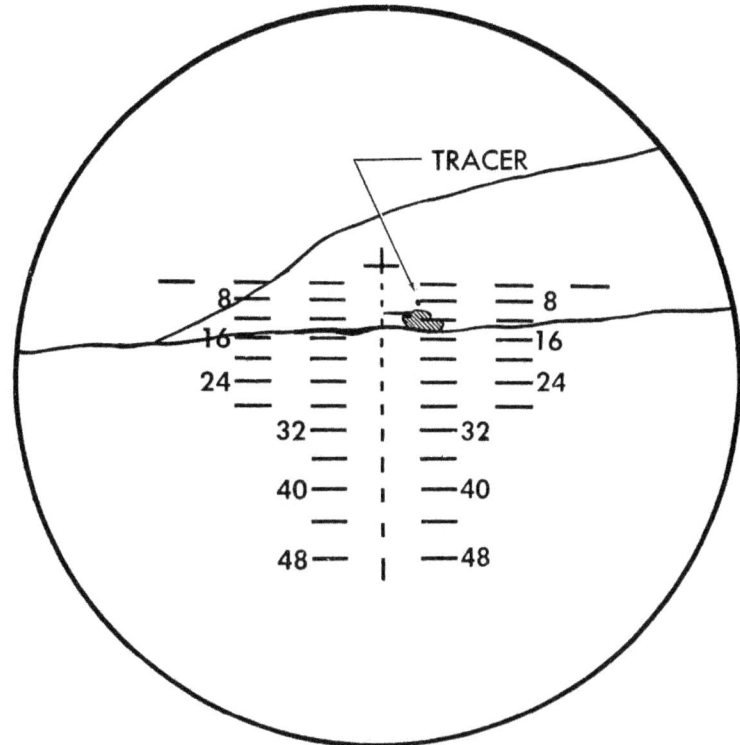

Figure 53. Situation 2—moving target.

elements of the subsequent fire command used when adjusting fire on moving targets is as follows:

Element	*Example*
(1) Range change (in yards).	ADD 200
(2) Lead change (in leads).	ONE HALF MORE
(3) Command to fire.	FIRE.

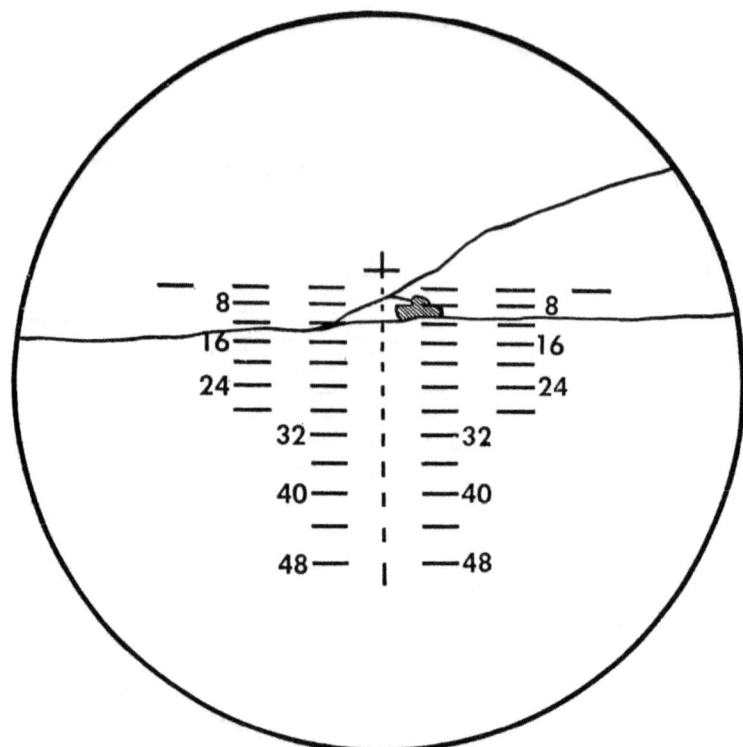

Figure 54. Situation 2—moving target.

Section V. FIRING TANK MACHINEGUNS

112. General

A tank develops its decisive ability when it closes with the enemy in the final assault, and it is in this phase that the tank machineguns play the major role. The machineguns may be fired from moving or stationary tanks, and the target engaged may be either moving or stationary. These guns furnish a great volume of fire, and a large supply of machinegun ammunition is carried in the tank. The tank gun crew fire the main gun only when the machinegun will not accomplish the mission. Therefore, the training of tank crewmen in the methods and techniques of firing the tank machineguns must be emphasized. The tank has two machineguns, the coaxial caliber .30 and the turret-mounted caliber .50. The technique in firing these machineguns is as follows:

a. Coaxial Machinegun. The commands used to engage either moving or stationary targets with the coaxial machinegun are given in the same sequence of elements as when firing the main gun (pars. 100 and 108). The coaxial machinegun is fired in bursts of 20 to 25 rounds; fire is adjusted by manipulating the tracer stream into and throughout the target area. An example of an initial fire command follows.

Element	Example
Alert.	GUNNER
Ammunition.	CALIBER THIRTY
Range.	FIVE HUNDRED
Direction.	(Omitted)
Description.	TRUCK
Lead.	ONE LEAD
Command to fire.	FIRE.

b. Turret-Mounted Machinegun. The turret-mounted machinegun is fired by the tank commander at ground or aerial targets. When engaging ground targets, he fires it in bursts of 10 to 20 rounds, adjusting fire by manipulating the tracer stream into and throughout the target area. Against aerial targets, he fires it in one continuous burst as long as the target is within range. By tracking, leading, and observing, he manipulates the tracer stream into the target.

Section VI. SPECIAL SITUATIONS

113. General

Situations arise in combat that prevent the use of normal techniques of fire and require that substitute means be utilized to insure target destruction. Examples of these conditions are—the destruction of dangerous surprise or rapidly fleeing targets, where the battle sight is employed; the destruction of dug-in and defiladed targets by ricochet fire; the destruction or neutralization of point and area type targets by use of massed fire; firing during periods of limited visibility with prepared range cards; and firing upon targets from a defiladed position.

114. Dangerous Surprise Targets

Any target capable of seriously damaging your tank (a tank, or an antitank or self-propelled gun), which has either fired at you or is capable of bringing fire to bear upon you immediately, is considered a dangerous surprise target and should be engaged with the battle sight (par. 116).

115. Fleeing Targets

Any target that is passing rapidly from view to take up a defiladed firing position or is about to escape destructions should be engaged with the battle sight. The tank commander will determine the priority of engaging targets and should not give priority to a target simply by virtue of its movement.

116. Battle Sight

A battle sight is a predetermined range setting combined with a particular type of ammunition. This range and ammunition setting will be determined by the unit commander as the most suitable combination to destroy the immediately dangerous surprise targets which are expected to appear in the area of combat operations. The two elements which compose the unit battle sight will vary according to information concerning the enemy, terrain, and weather. Normally, in combat operations, the main armament and the machineguns will be loaded. Unit commanders should include, in their standing operating procedure (SOP), the range and ammunition settings for battle sight. When employing battle sight, the following fire commands and techniques will be used.

a. Initial Fire Command—Stationary and Moving Targets.

Element	Example
Alert.	GUNNER
(Ammunition.	
(	BATTLE SIGHT
(Range.	
Description.	TANK
Leads.*	
Command to fire.	FIRE.

* Only when engaging moving targets.

(1) The ammunition and range elements will be omitted, since the battle sight includes a predetermined ammunition and range setting.

(2) The loader will continue to load the same type of ammunition until the tank commander changes ammunition or commands CEASE FIRE. If a change of ammunition is desired, the tank commander will announce FIRE HE, or any type of ammunition he desires to fire. The loader, upon hearing the word FIRE and a type of ammunition other than the type previously loaded, will select and continue to load the new type until the tank commander has again changed the ammunition element or commands CEASE FIRE.

Note. In the event that a round has been chambered and the ammunition element has been changed, the chambered round should be fired before the new type of ammunition is loaded.

b. Subsequent Fire Command—Stationary Targets. See paragraph 104.

c. Subsequent Fire Command—Moving Targets. See paragraph 111.

117. Ricochet Fire

The gunner cannot apply the primary method of adjustment when firing ricochet fire; therefore, the tank commander senses the effects of the fragments upon the ground and utilizes the alternate method of adjustment to bring the center of effect on the target. An example of an initial fire command for ricochet fire is as follows:

Element	Example
Alert.	GUNNER
Ammunition.	HE-DELAY
Range.	1,400
Description.	TROOPS
Command to fire.	FIRE.

118. Massed Fire

It is often desirable to mass the fire of two or more tanks against certain targets. Controlled concentrated fire by section, platoon, or company produces greater shock effect than does the uncoordinated fire of an equal number of tanks. Regardless of the situation, the target to be engaged by massed fire will be either a point target or an area target. Listed below are examples of initial fire comands to engage point and area type targets. Fire commands transmitted by radio must be preceded by call signs or other authorized unit designated. For additional information regarding techniques and procedures of massed fire, see FM 17-12.

a. Point Type Targets. A point target may be described as a lone building, a single antitank gun, a lone tank, or any single target that may be engaged by direct fire. The sequence of the elements is—

Element	Example
Alert.	PLATOON
Ammunition.	HE
Range.	MY RANGE 1,100*
Direction.	WATCH MY BURST
Description.	RED BRICK BUILDING
Command to fire.	FIRE.

*The range element is announced to assist the other tank commanders in locating the target designated by the burst of the fired round.

b. Area Type Targets. An area target is the concentration of several targets—for example, a large body of troops, a truck column, a concentration of enemy tanks, etc. The fire command to engage area targets is designed to insure complete target coverage.

(1) To engage a column of tanks from ambush, the following fire command would be used:

Element	Example
Alert.	PLATOON
Ammunition.	SHOT
Range.	MY RANGE 800
Direction.	DIRECT FRONT
Description.	NO. 2 LEAD TANK
	NO. 3 SECOND TANK
	NO. 4 FOURTH TANK
	NO. 5 LAST TANK
Command to fire.	FIRE.

(2) To engage an area target such as a large concentration of troops, the directional element assigns the sector of fire to the individual tanks. An example follows:

Element	Example
Alert.	PLATOON
Ammunition.	CALIBER THIRTY
Range	600*
Direction.	RIGHT FRONT
	NO. 2 RIGHT FLANK
	NO. 5 LEFT FLANK
	NO. 3 RIGHT CENTER
	NO. 4 LEFT CENTER
Description.	TROOPS
Command to fire.	FIRE.

* The range element is announced for location purposes. In this case, it designates the center of the sector.

119. Night Firing

To increase the effectiveness of tank weapons during the hours of darkness, it is imperative that commanders plan "on-call" artificial illumination. Artificial illumination is provided by illuminating shells, flares, fires, or searchlights. To deliver effective fire under artificial illumination, an instrument light must be used to illuminate the sight reticle. The rheostat on the instrument light enables the gunner to adjust its brightness. Too much light on the reticle blinds the gunner. For additional techniques on night firing, see FM 17-12.

120. Range Cards

a. A range card is a diagram or sketch of an area, showing the tank's position, the prominent terrain features, and the probable targets, all in their actual relation to positions on the ground. All objects shown on a range card are identified by description, range, quadrant elevation, and deflection (azimuth indicator reading).

b. During combat, range cards are constructed when the tanks are to be halted for any length of time. The platoon leader normally designates a reference point and the probable targets. Each tank crew then prepares a range card, recording the firing data to all the probable targets visible from the tank's position. The commander determines the range to the reference point by the most accurate means available. Next, the gunner indexes, on the ballistic unit, the appropriate range for HE ammunition, lays the aiming cross of the M20 periscope on the reference point, zeroes the azimuth indicator, and centers the bubble in the elevation quadrant. The range, deflection, and quadrant elevation are recorded. Data for all probable targets and areas in which targets may be expected to appear is obtained as follows:

    (1) The tank commander determines the range to the target by the most accurate means available.

    (2) The gunner indexes the appropriate ammunition and range on the ballistic unit and lays the aiming cross of the M20 periscope on the center of the probable target, using the manual turret controls.

    (3) Without disturbing the lay of the gun, the gunner centers the bubble in the elevation quadrant.

    (4) The commander then records the range, deflection, and quadrant elevation.

    c. For illustrations and further details on the use of the range cards, see FM 17-12.

121. Firing From Defilade

To engage targets from a defiladed position, the tank commander issues a six-element initial fire command in the sequence shown in paragraph 110. The range element is announced in hundreds of yards; and the gunner, referring to a firing table, determines the elevation for the announced range. He applies this to the elevation quadrant, centering the bubble by elevating or depressing the gun. The directional element of the fire command may be given by the reference point method, in which the tank commander lays the gun on a reference point and has the gunner traverse right or left, a given number of mils, using the azimuth indicator. The tank commander observes the effect of the fire and uses the alternate method of adjustment to obtain target hits. During firing from defilade, minimum elevation, minimum range, and angle of site must be considered. For details, see FM 17-12.

CHAPTER 6
TANK GUNNERY QUALIFICATION COURSE

Section I. INTRODUCTION

122. General

This chapter contains the prescribed Tank Gunnery Qualification Course and gunnery qualification standards for the M41 and M41A1 tanks.

123. Purpose and Scopes of Courses

a. The purpose of the Tank Gunnery Qualification Course is to provide a means of determining the proficiency of the tank crewman in gunnery. The course is designed both to test the gunner and to serve as an adjunct to training in the proper care and use of the weapons and their accessories. The tables of the course will be fired for training purposes before they are fired for record. The standard Tank Gunnery Qualification Course covers the Gunner's Preliminary Examination, subcaliber firing exercises, and service firing exercises; the limited course omits the service firing exercises. The tables fired in the two types are—

   (1) Standard: tables I through VIII.

   (2) Limited: tables I through IV.

b. Each tank crewman must pass the Gunner's Preliminary Examination with a score of 800 percent or higher before firing either the standard or the limited course. Gunners firing the standard course must attain a score of 280 or higher over tables I through IV before firing tables V through VIII. Since tank crewmen of the active Army must be considered ready for combat, without further training, after successfully completing the Tank Gunnery Qualification Course, a definite classification for such personnel, based only on a limited course, must not be made. Therefore, use of the limited course for classification purposes is restricted to tank crewmen of the reserve components. When ranges are not available to active Army personnel for service firing (tables V–VIII), the full limited course, practice and record firing, may be fired twice annually for practice; however, classification of gunners will not be made.

c. The total possible points that can be scored in each phase of qualification are as follows:

	Possible points
(1) Gunner's Preliminary Examination	320
(2) Subcaliber firing exercises (tables I–IV)	400
(3) Service firing exercises (tables V–VIII)	400

d. In order to obtain maximum use of facilities, tables may be fired in any order within each group, except that table I will be fired first in the subcaliber exercises, and table V will be fired first in the service firing exercises.

124. Classification of Gunners

a. Personnel successfully completing the standard course will be classified as expert gunners, first-class gunners, or second-class gunners.

b. Personnel successfully completing the limited course will be classified, subject to the limitation imposed by paragraph 123b, as limited first-class gunner or limited second-class gunner.

c. The required score for each classification is as follows:

Classification	Score
Standard course	
Expert gunner	720–800
First-class gunner	640–719
Second-class gunner	560–639
Unqualified	559 or less
Limited course	
Limited first-class gunner	360–400
Limited second-class gunner	320–359
Unqualified	319 or less

125. Ammunition Required

The following table lists the ammunition required to fire the Tank Gunnery Qualification Course.

Course (Fired once for practice and once for record)	Ammunition				
	Coaxial machine gun			Service	Service
	Tracer	Ball	Frangible	HE	Shot
Table I (1)	30				
II (2)	15				
III	30	120			
IV (2) (3)			15		
V (4)					4
VI (4)				4	4
VII (4)					10
VIII (4)				1	
Total practice	75	120	15	5 (4)	18 (4)
Total record			Same as for practice		

(1) Includes 10 rounds for zeroing and failure to fire single shot.
(2) Includes three rounds for zeroing.
(3) Tracer ammunition may be used if frangible ammunition is not available.
(4) Service firing not included in limited course.

126. Rules for Record Firing

a. Before record firing, the gunner is required to check the condition of the weapon, sights, and ammunition. He is permitted a reasonable length of time to do this.

b. Only the examining personnel, and the necessary personnel of the assigned crew of which the gunner being tested is a member, will be in the tank during record firing. The examiner will take the tank commander's position in the tank.

c. During firing, the gunner performs all the operations required by the test, without the benefit of coaching or assistance. Other members of the crew perform their normal duties. The assigned crewmen will rotate positions within their tank when firing the Tank Gunnery Qualification Course.

d. All exercises will be fired with the sights adjusted by boresighting and zeroing. (The established zero will be used for tables VI–VIII.)

e. If a misfire or other malfunction of the main gun occurs, the gunner will announce MISFIRE; if the machinegun fails to fire, the gunner will announce STOPPAGE. Thereafter, no one is allowed to touch the gun without authorization of the examiner. The examiner notes the time and examines the gun. After the malfunction has been corrected, the crewmen will be permitted to complete the exercise. Unless the malfunction was due to the negligence of the gunner, the time it takes to correct the malfunction will not be counted against the time allowed the gunner for that particular phase of the test.

f. Prior to record firing, the examiner must thoroughly familiarize himself with his duties and the correct firing procedure.

Section II. GUNNER'S PRELIMINARY EXAMINATION, GENERAL

127. General

The gunner's preliminary examination will be conducted by the company commander and such officer and enlisted assistants as may be necessary. The examination will be given to each member of the tank crew, and a score of 80 percent or more will be required before the crewman is permitted to fire the tables of the tank gunnery qualification course.

128. Table of Possible Scores

GUNNER'S PRELIMINARY EXAMINATION

	Possible points
Materiel tests	
Disassembly of breech mechanism	20
Assembly of breech mechanism	20
Care and maintenance	20
Sight adjustment	50
Putting turret into power operation	10
Testing gunner's quadrant	10
Adjusting elevation quadrant (omitted if tank is not equipped with elevation quadrant).	10
Identification and inspection of ammunition	10
Total possible score	150
Simulated firing tests	
Direct laying, primary sighting devices	40
Direct laying, secondary sighting devices	30
Use of elevation quadrant	30
Use of gunner's quadrant	30
Adjustment using azimuth indicator and elevation quadrant	40
Total possible score	170
TOTAL POSSIBLE SCORE, GUNNER'S PRELIMINARY EXAMINATION	320

Notes 1. A gunner must score 256 or more points to be eligible to fire subcaliber or service exercises.

2. For tanks not equipped with elevation quadrant, the total possible score for the Gunner's Preliminary Examination is 310 points, and a gunner must score 248 or more points to be eligible to fire subcaliber or service exercises.

Section III. MATERIEL TESTS, GUNNER'S PRELIMINARY EXAMINATION

129. Test on Disassembly of Breechblock

The breech cover is removed and the gun traveling lock disengaged. The examiner commands, DISASSEMBLE BREECHBLOCK, and starts his timing with a stop watch. The gunner is required to remove and disassemble the breechblock, using the prescribed method. If the disassembly is completed in 3 minutes or less, a credit of 20 points is given. A cut of 5 points is made for every additional 30 seconds or part thereof when more than 3 minutes are taken for this test. To facilitate the handling of parts in disassembly and assembly of the breechblock, all parts should be free from oil and grease. A tarpaulin or shelter half should be placed on the floor of the turret to avoid losing or damaging parts, and the turret lights should be turned on to improve visibility. The examiner will note any particular difficulty encountered by the gunner in removing or replacing any part. If the difficulty is due to the part being bent, burred, or otherwise damaged, the examiner will note the time required to remove or replace that part and deduct it from the total time required. The closing spring will *not* be removed in this test.

130. Test on Assembly of Breechblock

The examiner commands, ASSEMBLE BREECHBLOCK, and starts the timing with the stop watch. The gunner is required to assemble and replace the breechblock, using the prescribed procedure. If the assembly and replacement is correctly performed in 4 minutes or less, a credit of 20 points is given. A cut of 5 points is made for every additional 30 seconds or part thereof when more than 4 minutes are taken for this test. One other crewman (preferably an instructor) may assist the gunner when appropriate.

131. Test on Care and Maintenance

The examiner commands, PERFORM DAILY LUBRICATION CHECK, INCLUDING CHECKING AND FILLING OF RECOIL SYSTEM. The gunner points out the lubricating points for which he is responsible and, from a display of lubricants and lubricating devices, selects the proper ones to be used for each point. All lubricants will be in the containers in which they are normally issued. The gunner also explains the procedure for checking, filling, and bleeding the recoil system. The total possible credit is 20 points. A penalty of 4 points is assessed for each of the following errors:

    a. Each point of lubrication missed.

    b. Each error in selection of lubricating device or lubricant.

    c. Each error in procedure for checking, filling, and bleeding the recoil system.

132. Test on Sight Adjustment

The vehicle is placed in a position where several features or objects, suitable for sight adjustment, are in view. These objects should be at varied known ranges of from 500 to 3,000 yards. The gunner must select a target as near 1,500 yards as possible for boresighting. To conduct the exercise, the examiner places the gunner's M20 periscope, the ballistic unit, and the M97 telescope out of adjustment. He then commands, MAKE BORESIGHT ADJUSTMENT. The gunner is required to adjust the M20 periscope, the ballistic unit, and the M97 telescope, using the prescribed methods. The total possible credit is 50 points. A score of 40 points is given for placing the periscope and ballistic unit in proper adjustment, and 10 points for placing the telescope in proper adjustment. No partial credit will be given for items which are not properly adjusted.

133. Test on Putting the Turret Into Power Operation

The examiner commands, PUT TURRET INTO POWER. The gunner is required to put the power traverse mechanism into operation, performing all steps in the prescribed sequence. The total possible credit is 10 points. A penalty of 5 points is assessed for each error.

134. Test on Testing Gunner's Quadrant

The examiner commands, TEST GUNNER'S QUADRANT. The gunner is required to make the end-for-end test on a quadrant which is out of adjustment. The gunner is also required to explain the method of determining the correction. The total possible credit is 10 points. A penalty of 5 points is assessed for each error.

135. Test on Adjusting Elevation Quadrant

(Omitted if tank is not equipped with elevation quadrant.) The elevation scale and the micrometer scale of the elevation quadrant are placed out of adjustment. The examiner commands, ADJUST ELEVATION QUADRANT. The gunner is required to adjust both the elevation and the micrometer scales, using the prescribed procedure. No credit is given if the adjustments are not precise. The total possible credit is 10 points.

136. Test on Identification and Inspection of Ammunition

All standard types of ammunition, for both the tank gun and the machinegun, are displayed. Some rounds will have apparent faults, such as a dented cartridge case, badly burred rotating band, or adhering dirt. Lettering and markings on service ammunition are covered. The gunner is required to identify five rounds of ammunition as they are pointed out to him. Color, shape, and size are used as means of identification. The gunner is also required to locate the defects in three rounds that are pointed out to him, and to describe the results of using these rounds without correcting the defect. The total possible credit is 10 points. A penalty of 2 points is assessed for each error.

Section IV. SIMULATED FIRING TESTS, GUNNER'S PRELIMINARY EXAMINATION

137. Test on Direct Laying, Primary Sighting Devices (M20 Periscope and M4 Ballistic Drive)

In this test, the gunner sets off on the ballistic unit the correct range for the ammunition announced by the examiner and lays the gun with the correct sight picture on four targets.

a. The tank is placed in a position from which several targets can be seen. (The aiming point on each target must be well defined to eliminate confusion as to the correct laying of the gun.) The gunner checks the diopter setting on the M20 periscope, sets the unit battle sight on the ballistic unit, and announces READY.

b. The examiner gives a five-element initial fire command while laying the gun for direction with the commander's power control handle and M20 periscope. The examiner starts timing and the gunner starts the test (sets range on ballistic unit) when the range is announced. The gunner makes final lay of the gun on the target and announces ON THE WAY. The examiner checks the initial sight picture, using the commander's M20 periscope, and checks the ballistic unit to determine whether the correct range has been set for the announced ammunition. If the gunner performs this exercise correctly within 10 seconds, he receives 10 points. He will receive no credit if he does not perform the exercise correctly within 10 seconds.

c. The test is then repeated three times, using the procedure outlined in *b* above, for a total of four trials.

d. The total possible score for this test is 40 points.

138. Test on Direct Laying, Secondary Sighting Devices (M97 Telescope and M4 Ballistic Drive)

In this test, the gunner sets off on the ballistic unit the correct range for the ammunition announced by the examiner and lays the gun for the correct initial sight picture. Then, on command, the gunner applies two subsequent corrections to the gun by use of the sight reticle. The ammunition element announced by the examiner will be other than that for which the sight is graduated. The crewman being tested will use the ballistc unit or an aiming data chart to determine the correct range to apply with the telescope.

Note. If the ammunition element of the initial fire command is other than that for which the sight is graduated, the ballistic unit or an aiming data chart must be used.

a. The tank is placed in a position from which several targets can be seen. (The aiming point on the target must be well defined to eliminate confusion as to the correct laying of the gun.) The gunner checks the diopter setting on the M97 telescope, sets the unit battle sights on the M4 ballistic unit, and announces READY.

b. The examiner gives a five-element initial fire command while laying the gun for direction with the commander's power control handle and M20 periscope. The examiner starts timing and the gunner starts the test (sets range on ballistic unit) when the range is announced. The gunner, using the ballistic unit or an aiming data chart, determines the correct range to use on the telescope, makes the final lay of the gun, and announces ON THE WAY. The examiner checks the initial sight picture, using the gunner's telescope, and checks the ballistic unit to determine whether the correct range has been set opposite the announced ammunition. If the gunner performs this exercise correctly within 10 seconds, he receives 10 points. He will receve no credit if he does not perform the exercise correctly within 10 seconds.

c. The examiner then gives a subsequent fire command, using a deflection change of not more than 10 mils and a range change of not more than 400 yards. The gunner lays the gun, applying the announced correction by using the sight reticle, and announces ON THE WAY. The examiner checks the sight picture, using the gunner's telescope. If the gunner performs this exercise correctly within 5 seconds, he receives 10 points. He will receive no credit unless he has the correct sight picture within 5 seconds.

d. The examiner then gives a second subsequent command, using the procedure outlined in *c* above.

e. The total possible score for this test is 30 points.

139. Test on Use of Elevation Quadrant

In this test, the gunner lays the gun for range, using the elevation quadrant (or Gunner's Quadrant, M1, on a tank without elevation quadrant).

a. The examiner announces a range in yards. Using the ballistic unit as an aiming data chart, the gunner determines the elevation corresponding to the announced range, insures that the ballistic unit is returned to zero, applies this elevation to the elevation quadrant (or M1 quadrant if the tank is not equipped with an elevation quadrant), and centers the bubble, using the turret controls. He then announces ON THE WAY. The examiner checks the elevation setting and the correct lay of the gun.

b. The examiner then announces two subsequent ranges, and the same procedure is followed.

c. No credit will be given if an improper quadrant setting is used or if the bubble is not accurately centered. For each trial, if the trial is correctly performed in exactly 9 seconds or less (12 seconds or less if the M1 quadrant is used), the gunner receives 10 points. He will receive no credit if the time exceeds 9 (12) seconds. The total possible score for this test is 30 points.

140. Test on Use of Gunner's Quadrant, M1

In this test the gunner sets an elevation on the M1 quadrant and places the quadrant properly on the quadrant seats.

a. The gunner sets the M1 quadrant at zero. The examiner announces an elevation. The gunner applies this elevation, properly seats the quadrant on the breech, and announces SET. The examiner checks the elevation and the position of the quadrant.

b. The examiner then announces two subsequent elevations, one of which will be to an even tenth of a mil, and the same procedure is followed.

c. No credit will be given if the elevation is incorrect or if the quadrant is seated improperly. For each trial, if the trial is correctly performed in exactly 7 seconds or less, the gunner receives 10 points. He will receive no credit if the time exceeds 7 seconds. The total possible score for this test is 30 points.

141. Test on Adjustment Using the Azimuth Indicator and the Elevation Quadrant

On tanks that are not equipped with an elevation quadrant, use the M1 gunner's quadrant.

a. The examiner sets the azimuth indicator at zero and the elevation quadrant at an arbitrary elevation, centering the bubble with the turret controls. This takes the place of laying the gun from an initial fire command.

b. The examiner then gives four subsequent commands. The deflection shifts should be in multiples of 5 mils and should not exceed 100 mils. The gunner follows the commands, using the azimuth indicator for deflection changes and the elevation quadrant for range changes. After each azimuth indicator and quadrant setting has been applied and the bubble centered, the gunner announces ON THE WAY. The examiner checks the azimuth indicator and quadrant settings and the bubble of the elevation quadrant.

c. No credit will be given if any setting is incorrect or if the bubble of the quadrant is not centered. For each trial correctly performed in exactly 20 seconds or less, the gunner receives 10 points. He will receive no credit if the time exceeds 20 seconds.

(If the M1 gunner's quadrant is used, allow 30 seconds.) The total possible score for this test is 40 points.

Section V. SUBCALIBER FIRING EXERCISES

142. General

All subcaliber firing will be conducted with the coaxial machinegun. When single shots must be fired, ammunition must be loaded with alternate dummy rounds, or a single-shot device may be used. Controller, Single-shot, Caliber .30-Caliber .50, or CONARC- approved training aid No. 1, single-shot device, may be used. Targets will be physically scored during all record firing.

143. Table I: Subcaliber Manipulation Exercise, 1,000-Inch

a. The purpose of this exercise is to test the gunner's ability to manipulate the turret controls and to fire at stationary targets prior to firing service ammunition.

b. This exercise in manipulation requires the gunner to fire rapidly on a series of stationary targets; figure 55 illustrates the target layout. The 4 x 4-inch targets shown in the illustration may be stapled or tacked onto staves.

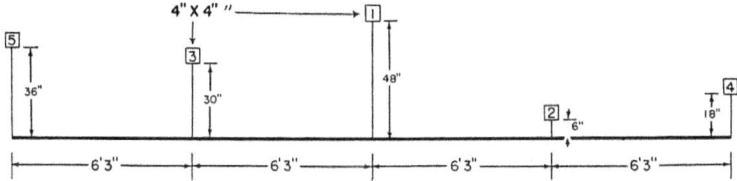

Figure 55. Manipulation target layout.

c. The coaxial machinegun is zeroed for 1,000-inch firing, using the telescope sight. After zeroing, the gunner lays on the left (No. 5) target. At the command, COMMENCE FIRING, he lays the aiming point of the sight reticle on the center (No. 1) target and fires one round. He then fires one round at each of the remaining four targets in the order in which they are numbered. Time is recorded from the command, COMMENCE FIRING. At the end of the allowed time, the examiner will command, CEASE FIRING. Rounds fired after this command will be scored as misses.

d. The exercise consists of four trials as outlined in *c* above, two in manual traverse and two in power traverse. Credit will be given in accordance with table I.

e. The exercise will be scored as follows:

Table I. Possible Score: 100

Trials	Number rounds	Time (seconds)	Points each hit
1st—Manual traverse	5	30	5
2d —Manual traverse	5	30	5
3d —Power traverse	5	40	5
4th—Power traverse	5	40	5

Note. Targets will be scored and marked, and replaced when required.

f. See figure 56 for illustration of an appropriate score card.

Co __A__
Bn __1ST__

NAME __SMITH JOHN J.__
RANK __PFC__ SN __38132977__
DATE __1 FEB 54__

__75__
Total Score

TANK GUNNERY QUALIFICATION COURSE SCORE CARD

100 points Possible

TABLE I (SUBCALIBER MANIPULATION EXERCISE—1000-INCH)

| TRIALS | NUMBER OF ROUNDS | POSSIBLE | MAX TIME (SEC) | TARGET HITS | | | | | SCORE |
				1st RD	2d RD	3d RD	4th RD	5th RD	
1st—Manual	5	25	30	5	5	5	0	0	15
2d—Manual	5	25	30	5	5	0	5	5	20
3d—Power	5	25	40	5	5	5	5	0	20
4th—Power	5	25	40	5	5	0	5	5	20
								TOTAL SCORE	75

A. Five points for each target hit. No credit for hits obtained after time limit.
B. Add score column to obtain total score for this exercise.

Satisfactory Score _____ 70 points.

Lt Leonidas K. James
Examining Officer's Signature

Figure 56. Score card for table I.

144. Table II: Subcaliber Shot Adjustment, Moving Target Exercise, 200 Feet

a. The purpose of this exercise is to test the gunner's ability to lead, track, engage, and adjust fire on moving targets prior to firing service ammunition.

b. In this exercise, the gunner is required to fire tracer (single shot) at moving targets on a 200-foot range, using the coaxial machinegun.

c. Targets are mounted on a 6 x 6-foot panel as shown in figure 57. The target shown is adequate to fire three tanks simultaneously; if it is desired to fire more tanks, targets may be pulled in tandem. The speed of the targets should be approximately 5 miles per hour.

d. When a moving target range is not available, target tanks may be utilized for this exercise if available and desirable. Ranges for target tanks should be about 400 yards.

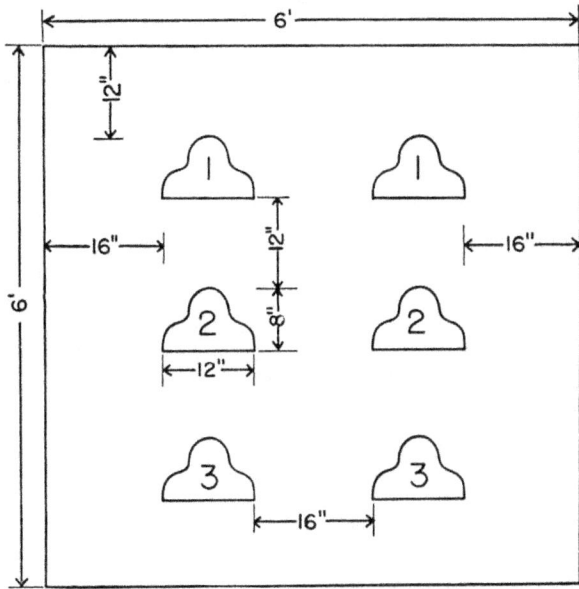

Figure 57. Moving target layout.

e. Using the M20 periscope, with a range of 800 yards for shot ammunition indexed on the ballistic unit, the coaxial machinegun is zeroed to hit the center of the target at a distance of 200 feet when one lead is taken. If the periscope will not converge with the coaxial machinegun at a distance of 200 feet, the telescope will be used and zeroed to hit the center of the target with a sight picture of 800 yards when one lead is taken; however, every effort should be made to zero the primary sight. A target traveling 5 miles per hour requires only one half of a lead. The gun, therefore, must be zeroed on a stationary target with one half lead in order for the gunner being tested to hit the moving target with a sight

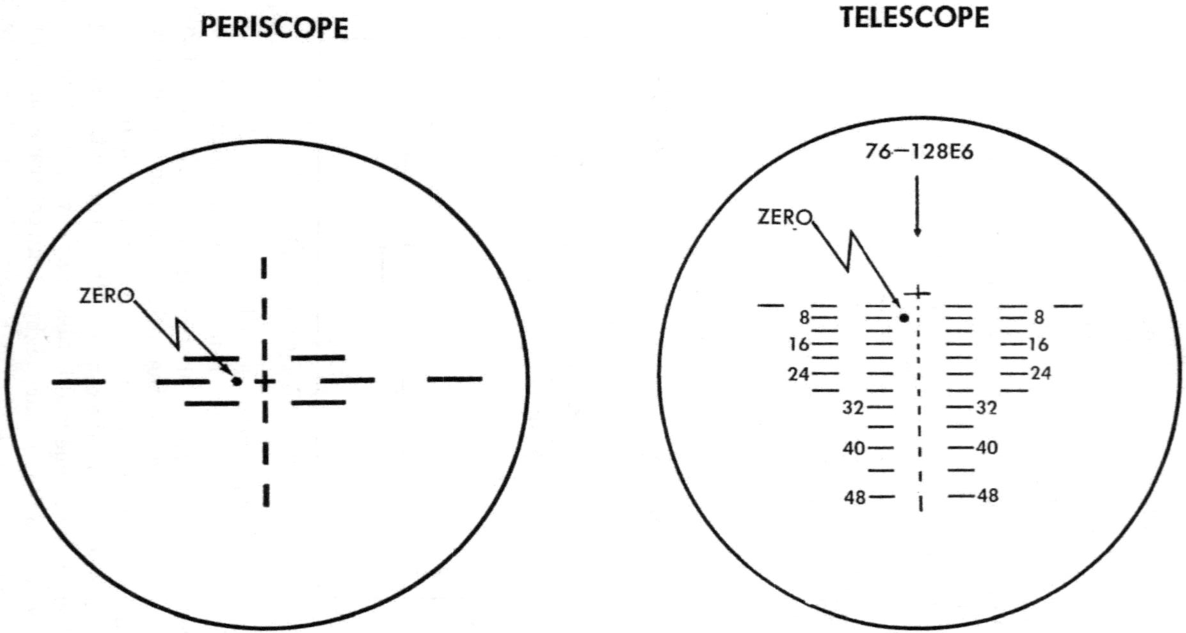

Figure 58. Zeroing for table II.

picture of one lead. Figure 58 shows the proper zero for both the primary and secondary sights to fire on a target traveling from left to right.

f. The exercise is conducted as follows:

(1) As the targets move along the course, the examiner gives a fire command announcing a range of 800 yards and designating the target by number. The gunner then fires three rounds at the designated target. The examiner records time from the command FIRE until after the third round is fired.

(2) The exercise consists of four trials as outlined in (1) above; two trials are fired at the lead target in manual traverse and two at the rear target in power traverse.

g. The exercise will be scored as follows:

Table II. Possible Score: 100

Trials	Number rounds	Possible	Minus	Score
Manual	3	25		
Manual	3	25		
Power	3	25		
Power	3	25		

Cuts for each target:
 Failure to fire first round within 5 seconds _____ 5 points.
 Failure to complete trial in 15 seconds _____ 5 points.
 Each round that fails to hit target _____ 5 points.

h. See figure 59 for illustration of an appropriate score card.

145. Table III: Moving Tank Exercise (Stationary Target)

a. The purpose of this exercise is to test the gunner's ability to fire the coaxial machinegun from a moving tank at stationary targets.

b. In this exercise, the gunner fires 150 rounds from a moving tank at groups of targets representing infantry.

c. The length of the course (fig. 60) will be approximately 800 yards. The end of the course will be marked by two white flags, one on each side of the runway. Targets will be kneeling-type (E) silhouettes. Five groups of four silhouettes will be placed not more than 10 yards nor less than 5 yards from the sides of the tank runway, alternately on the left and right sides. One of these groups will be placed at each of the following ranges from the starting point: 200 yards, 350 yards, 450 yards, 550 yards, and 700 yards. A sixth group of five silhouettes will be placed 200 yards beyond the end of the course and in direct line with the center of the

Co. **A**
Bn. **1ST**

NAME **SMITH JOHN J.**
RANK **PFC** SN **38132977**
DATE **2 FEB 54**

90
Total Score

TANK GUNNERY QUALIFICATION COURSE SCORE CARD

100 points Possible

TABLE II (SUBCALIBER SHOT ADJUSTMENT MOVING TARGET EXERCISE—200 FEET)

TRIALS	NUMBER OF ROUNDS	POS-SIBLE	1ST ROUND FIRED IN 5 SEC		TRIAL COMPLETED IN 15 SEC		TARGET HITS			SCORE
			YES	NO	YES	NO	1ST RD	2D RD	3D RD	
1st—Manual	3	25	X		X		0	5	5	20
2d—Manual	3	25	X		X		5	5	5	25
3d—Power	3	25	X		X		5	6	0	20
4th—Power	3	25	X		X		6	6	5	25
									TOTAL SCORE	90

A. Five points for each "Yes."
B. Five points for each Hit.

Satisfactory Score ------------------------------------- 70 points.

Lt Leonidas N. James
Examining Officer's Signature

Figure 59. Score card for table II.

course. Each group of silhouettes will cover an area 4 yards wide and 4 yards deep. A red flag will be placed on the edge of the runway, 50 yards from each silhoutte group, in the direction of the starting line.

   d. The terrain selected for the course will be such that the tank can maintain an average speed of 5 miles per hour.

   e. The ammunition will be loaded four ball to one tracer.

   f. The test is conducted as follows:

      (1) The gunner will make only one run over the course while being tested. Fire will cease on a target group when the tank reaches the red flag 50 yards from that group. All

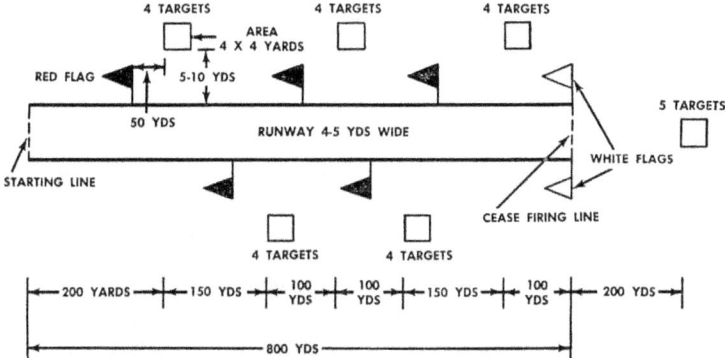

Figure 60. Range setup—moving tank exercise.

hatches except the tank commander's will be closed at the starting line and will remain closed throughout the run.

(2) The tank will not stop until it reaches the white flags. At this point, firing will cease and the gun will be cleared. The tank will be required to maintain an average speed of 5 miles per hour between the starting line and the white flags.

(3) Separate courses may be set up and used concurrently, provided they are at least 30 yards apart.

(4) The examining officer will follow behind the tank in a vehicle and control its movement by radio. Assistants will follow to mark targets and score.

(5) An assistant will check the number of rounds in each belt before the tank begins the run.

g. The exercise will be scored as follows:

Table III. Possible Score: 100 Points

Target groups	Number rounds	Possible	Minus	Score
1	25	16 points		
2	25	16 points		
3	25	16 points		
4	25	16 points		
5	25	16 points		
6	25	20 points		
Total		100 points		

(1) Four points will be awarded for each silhouette that is hit regardless of the number of hits in each target. The maximum score is 100 points.

(2) No credit will be given for hits on a target group if that target group was fired on after the tank passed the red flag for that group.

(3) The gunner receives no credit for the course if his tank fails to sustain a 5-mile-per-hour speed, but he will be retested.

h. See figure 61 for illustration of an appropriate score card.

Co *A*
Bn *1ST*

NAME *SMITH JOHN J.*
RANK *PFC* SN *38132977*
DATE *3 FEB 54*

80
Total Score

TANK GUNNERY QUALIFICATION COURSE SCORE CARD

100 points
Possible

TABLE III (MOVING TANK EXERCISE, STATIONARY TARGETS)

TARGET GROUPS	NUMBER OF ROUNDS	POSSIBLE	TARGET HITS					SCORE
			NR 1	NR 2	NR 3	NR 4	NR 5	
1	25	16	4	4	0	4		12
2	25	16	4	0	4	4	(Applies to target group Nr. 6 only)	12
3	25	16	4	4	4	4		16
4	25	16	0	4	4	4		12
5	25	16	4	4	4	0		12
6	25	20	4	4	4	0	4	16
							TOTAL SCORE	80

A. Four points for each target hit. (No credit for target hits after passing red flag of each group.)

B. No credit if tank does not maintain a speed of 5 mph.

Satisfactory Score _____ 70 points.

Lt Leonidas K. James
Examining Officer's Signature

Figure 61. Score card for table III.

146. **Table IV: Auxiliary Fire Control Exercise**

a. Purpose. The purpose of this exercise is to test the ability of the gunner in the proper use of the tank's auxiliary fire-control instruments.

b. *Target Layout.*
  (1) The impact area selected should be fairly flat, and the surface should be of dirt or sand.
  (2) An aiming post, to serve as a gun reference point and 1,000-yard range marker, is placed in the center of the tank's impact area at a distance of 117 feet from the muzzle of the coaxial machinegun.
  (3) At varying simulated ranges (500–1,500 yards), small vertical targets (4 x 6-inch cards) will be fastened to the ground. The bottom edges of the targets will be folded to provide a fastening surface and a 4 x 4-inch target facing the tank. (See fig. 62 for target layout.)

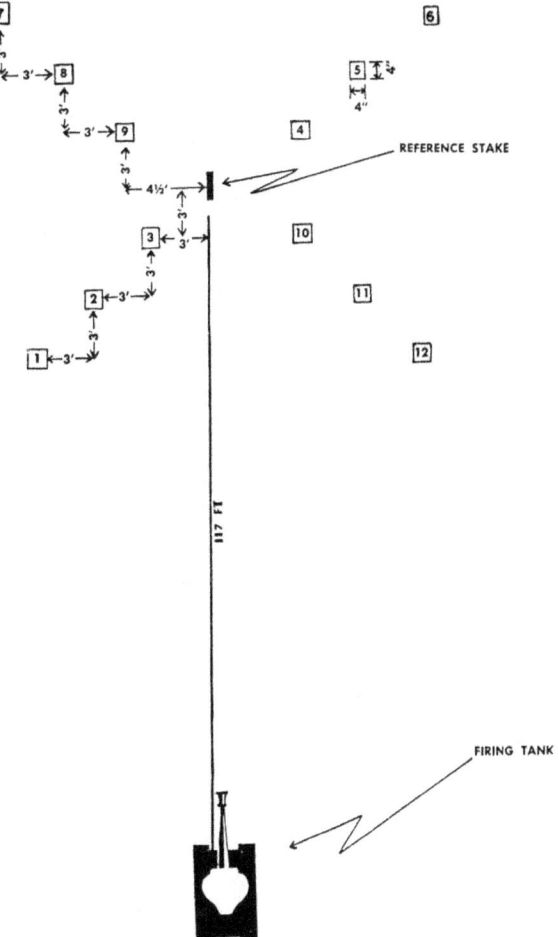

Figure 62. Target layout, table IV.

c. Preparation by the Examiner.
  (1) With the coaxial machinegun properly mounted and adjusted, the examiner fires and adjusts until he obtains a hit on the base of the gun reference point.
   Note. A 4 x 4-inch target is placed immediately behind the aiming post as an aid in zeroing.
  (2) Without disturbing the lay of the gun, the examiner will—
   (*a*) Center the bubble of the elevation quadrant with the micrometer knob. (On tanks not equipped with an elevation quadrant, use the M1 gunner's quadrant on the gun recoil guard. An inclined firing line with the front of the tank lower than the rear will allow the use of the M1 gunner's quadrant on the quadrant seats.)
   (*b*) Loosen the micrometer knob and slip the scales until a reading of 50 mils appears opposite the index, then tighten the micrometer knob.
   (*c*) Place a fine chalk or pencil line on the elevation scales opposite the index. This is to avoid 100-mil errors when the gun is elevated between problems. (On tanks where the M1 gunner's quadrant is used, the setting to hit the reference point will be recorded and will be the reference mil reading.)
   (*d*) Zero the azimuth indicator.
   (*e*) Adjust the boresight cross of the telescopic sight (using the boresight knobs) to the point of strike on the gun reference point.
  (3) After zeroing on the reference point, the examiner will determine the azimuth indicator reading and elevation quadrant reading to the center of each target in the impact area and will record this data on a reference card (fig. 63). The elevations are converted to ranges, which are also recorded on the reference card.
   Note. Make certain it is clearly understood that this is a REFERENCE card and *not* a RANGE card.
   Note. Elevations are converted to ranges by reference to the following: The elevation for a range of 1,000 yards equals 50 mils except for tanks not equipped with elevation quadrant. A 100-yard range change is effected by making a 1-mil change in elevation.
  (4) The gunner will use a 1-mil change in elevation for each 100-yard range change desired.

d. Procedure for Testing.
  (1) The gunner will be required to engage four targets, using the auxiliary fire-control instruments. The direct-fire sights will be covered during the exercise. Three rounds

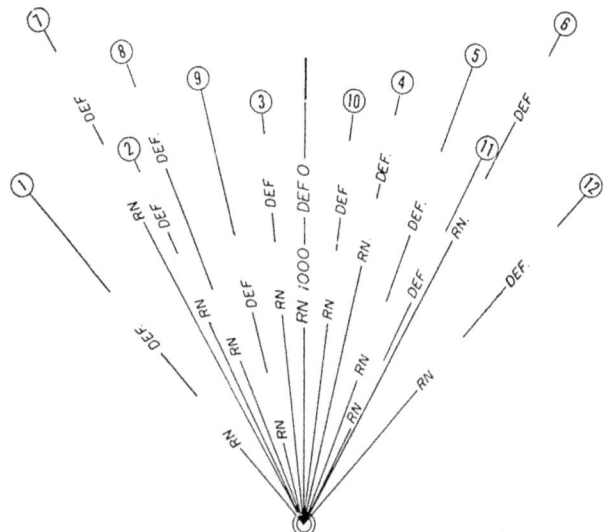

Figure 63. Reference card for table IV.

of caliber .30 frangible ammunition will be fired at each target; tracer ammunition may be used if caliber .30 frangible ammunition is not available. (The single-shot device can be used, or dummy rounds can be inserted between live rounds to permit firing single shots.)

(2) The examiner, acting as tank commander, will issue correct initial fire commands, sense each round, and give the necessary subsequent commands for each target. For example:
GUNNER
HE
1,500 (Correct range to target.)
FROM REFERENCE POINT RIGHT 30 (This will be based on previously computed data.)
ANTITANK
FIRE. (Time will be recorded from this command.)
Subsequent commands will be in accordance with the alternate method of adjustment. Time for subsequent rounds will also be recorded from the command FIRE.

(3) The gunner, in executing these commands, will use the specially prepared firing table and the elevation quadrant for initial and subsequent ranges.

(4) In making initial and subsequent deflection shifts, the gunner will use the manual traversing control and the azimuth indicator. At the conclusion of firing on each

target, the gun is returned to the reference point and the bubble in the elevation quadrant is centered with a reading of 50 mils.

Co __A__
Bn __1st__

NAME __SMITH, JOHN J.__
RANK __PFC__ SN __38132977__
DATE __3 FEB 54__

__80__
Total Score

TANK GUNNERY QUALIFICATION COURSE SCORE CARD

100 points
Possible

TABLE IV (AUXILIARY FIRE CONTROL EXERCISE)

TRIALS	Nr OF RDS	POS-SIBLE	1ST RD IN 20 SEC		2D RD IN 10 SEC		3D RD IN 10 SEC		TARGET HITS			CUTS	SCORE
			YES	NO	YES	NO	YES	NO	1ST RD	2D RD	3D RD		
Target 1	3	25		X	X		X		O	X	X	5	20
Target 2	3	25	X		X		X		O	O	X	5	20
Target 3	3	25	X		X			X	X	O	O	10	15
Target 4	3	25	X		X		X		X	X	O	O	25

TOTAL SCORE 80

A. CUTS: Cut five points for each "NO."
B. Cut five points for each target miss over one (maximum 10 points).

Satisfactory Score _____ 70 points.

Lt Lemida K. Jones
Examining Officer's Signature

Figure 64. Score card for table IV.

e. Scoring. Scoring for each problem will be conducted as follows:

Table IV. Possible Score: 100 Points

Trial	Number rounds	Possible	Minus	Score
Target 1	3	25 points		
Target 2	3	25 points		
Target 3	3	25 points		
Target 4	3	25 points		
Total		100 points		

Cuts for each target:
 Failure to get first round off in 20* seconds_____ 5 points.
 Failure to get second round off in 10* seconds_____ 5 points.
 Failure to get third round off in 10* seconds_____ 5 points.
 For each target miss over one_____ 5 points.

* On tanks that are not equipped with elevation quadrant, allow 30 seconds for each round.

f. Score Card. See figure 64 for illustration of an appropriate score card.

Section VI. SERVICE FIRING EXERCISES

147. General

The service ammunition tables are fired only after the gunner has qualified in tables I through IV. Ammunition for both practice and record firing should be of the same lot. Targets will be accurately scored during all record firing; BC scopes or similar viewing instruments may be used for this purpose.

148. Table V: Zeroing Exercise, Service Firing

a. General. This exercise is designed to test the gunner's ability to zero the M20 periscope and the M97 telescope, using shot ammunition on a stationary 6 x 6-foot panel target with a well-defined aiming point (fig. 65).

b. Boresighting.
  (1) The gunner is required to boresight the M20 periscope and the M97 telescope on the zeroing target, which should be at a range as near 1,500 yards as possible.
  (2) After boresighting, the gunner will lay the aiming cross on the zeroing target.

c. Zeroing.
  (1) The gunner indexes, on the ballistic unit, the correct range to the target for the type of ammunition being fired. With

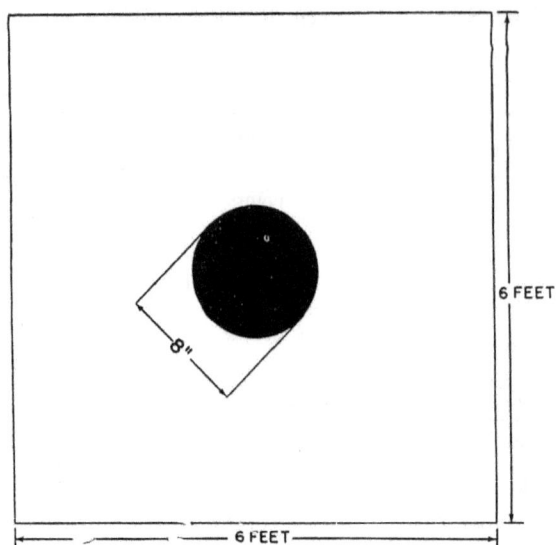

Figure 65. Target panel for zeroing exercise.

the aiming cross of the gun-laying reticle on the aiming point (center of 8-inch bull's-eye), he fires three rounds to form a shot group, checking after each round and relaying if necessary.

(2) After obtaining a group, the gunner re-lays on the aiming point and, without disturbing the lay of the gun, adjusts the aiming cross of the periscope and the appropriate range line of the telescope to the center of the group.

(3) The gunner re-lays on the aiming point and fires one check round to determine whether he has zeroed correctly.

d. Scoring. There is no time limit for this exercise. Scoring will be accomplished in accordance with table V.

Table V. Possible Score: 100 Points

No.	Item	Possible	Minus	Score

1	Correct bore sight procedure, M20 periscope and M97 telescope__Total value__30 points
	    For improper selection of borsight point, cut_____ 6 points
	    For failure to remove all superelevation prior to boresighting, cut_____ 6 points
	    For inaccuracy in laying aiming crosses of reticles on boresighting points observed through bore, cut_____ 6 points
	    For failure to lock boresight locking knobs after placing aiming crosses on boresighting point, cut_____ 6 points
	    For failure to set the slip scales properly on the boresight knobs of the M20 periscope, cut_____ 6 points			

2	Correct zeroing procedures, M20 periscope and M97 telescope__Total value__30 points
	    For failure to index correct range opposite proper type ammunition on ballistic unit, cut_____ 10 points
	    For failure to place proper range line of M97 telescope on center of shot group, cut_____ 10 points
	    For failure to use boresight knobs in manipulating aiming point of reticles on center of shot group, cut_____ 10 points			

3	Accuracy of zero_____Total value__40 points
	If check round strikes:
	    Within 14 inches from center of aiming point, no cut.
	    More than 14 and less than 18 inches from center of aiming point, cut____ 10 points
	    More than 18 and less than 24 inches from center of aiming point, cut____ 20 points
	    More than 24 inches from center of aiming point, cut_____ 40 points			

Note. A physical check of the target will be made to insure positive scoring of the check round.

e. Score Card. See figure 66 for illustration of an appropriate score card.

149. Table VI: Service Firing Exercise, Stationary Targets at Variable Ranges (Shot and HE Adjustment)

   a. This exercise is designed to test the gunner's ability to utilize

Co __A__   NAME __SMITH, JOHN J.__
Bn __1ST__   RANK __Pfc__ SN __38132977__
             DATE __4 FEB 1954__

__84__
Total Score

TANK GUNNERY QUALIFICATION COURSE SCORE CARD

100 points Possible

TABLE V (ZEROING EXERCISE FOR M41 AND M41A1 TANKS)

NR 1	BORESIGHTING	YES	NO	SCORE
(30 points) 6 points for each YES.	1. Properly selected boresight point.	X		6
	2. Removed all superelevation prior to boresighting.	X		6
	3. Aiming crosses of reticles accurately laid on boresighting point observed through bore.	X		6
	4. Boresight knobs locked after placing aiming crosses on boresighting point.	X		6
	5. Slip scales properly set on the boresight knobs of the M20 periscope.		X	
Nr 2	ZEROING PROCEDURE			
(30 points) 10 points for each YES.	1. Correct range indexed opposite proper type of ammunition on ballistic unit.	X		10
	2. Proper range line on M97 telescope placed on center of shot group.	X		10
	3. Manipulating aiming points of reticles on center of shot group by use of boresight knobs.	X		10
NR 3	ACCURACY OF ZERO			
40 points for YES.	Check round within 14 inches from the center of aiming point.		X	
30 points for YES.	Check round more than 14 and less than 18 inches from the center of aiming point.	X		30
20 points for YES.	Check round more than 18 and less than 24 inches from center of aiming point.		X	
			TOTAL	84

Satisfactory Score _____ 70 points

Lt Leonidas K. James
Examining Officer's Signature

Figur 66. Score card for table V.

the primary sighting equipment and the burst-on-target method of adjustment while firing service ammunition at stationary targets. The exercise is more realistic if tanks move between problems. The gunner will fire at four separate targets (two shot and two HE). The initial range to the target will be given by the examiner, and, if necessary, the gunner will apply the burst-on-target method of adjustment for the second round. The gunner will be limited to two rounds of ammunition for each of the four targets he is to engage.

b. The examiner will indicate each of the targets by issuing an initial fire command. The examiner will lay the gun for direction, using the commander's power control handle and the M20 periscope. Time for each problem starts when the command FIRE is announced in the initial fire command.

c. If the first round is not a target hit, the gunner will use the burst-on-target method of adjustment to fire the second round. If the target is hit on the first round, full credit will be given and the second round will not be fired.

d. Targets to be used for this exercise are 3 x 5-foot cloth panels for targets 1 and 2, and 6 x 6-foot cloth panels for targets 3 and 4. One HE and one shot problem will be fired at each size target.

e. Scoring for this exercise will be conducted as indicated in table VI.

Table VI. Possible Score: 100 Points

Trial	Range	Number rounds	Possible	Minus	Score
Target 1	800–1,100 yards	2	25		
Target 2	1,100–1,500 yards	2	25		
Target 3	1,500–1,800 yards	2	25		
Target 4	1,800–2,000 yards	2	25		

Cuts for each target:
 Failure to fire first round within 15 seconds _____ 5 points.
 Deduct one point for each additional second over
  15 required to fire first round, up to 20 seconds _____ 5 points.
 Failure to hit target with first round _____ 10 points.
 Failure to hit target with second round, if fired _____ 5 points.

f. See figure 67 for illustration of an appropriate score card.

150. Table VII: Moving Target Exercise (Shot)

a. The purpose of this exercise is to test the ability of the gunner to fire on moving targets. The gunner fires at five moving targets, using the burst-on-target method of adjustment for the second round fired at each target. The gunner will be limited to two rounds of ammunition for each of the five targets he is to engage.

Co __A__
Bn __1ST__

NAME __SMITH, JOHN U.__
RANK __PFC__ SN __38132977__
DATE __4 FEB 54__

POSSIBLE SCORE __100__
TOTAL CUTS __28__
TOTAL SCORE __72__

TANK GUNNERY QUALIFICATION COURSE SCORE CARD

TABLE VI (SERVICE FIRING, HE AND SHOT ADJUSTMENT).

	A			B		C	
TRIAL	NUMBER OF ROUNDS	POSSIBLE POINTS	TIME 1ST ROUND FIRED (Seconds)	TARGET HIT 1ST ROUND		TARGET HIT 2D ROUND	
				YES	NO	YES	NO
TARGET 1	2	25	12		X		
2	2	25	18		X		
3	2	25	11			X	X
4	2	25	9			X	X

Cuts:

		MAXIMUM CUTS	TARGETS				TOTAL CUTS
			1	2	3	4	
A. Failure to fire 1st round within 15 seconds	5 points		0	5	0	0	5
(Deduct one point for each second over 15.) Maximum cut.	5 points		0	3	0	0	3
B. Failure to hit target with 1st round.	10 points		0	0	10	10	20
C. Failure to hit target with 2d round.	5 points		0	0	0	0	0
(In case the target is hit by 1st round the 2d round will not be fired.)					TOTAL		28

Satisfactory Score _____ 70 points.

Lt Leonidas V. James
Examining Officer's Signature

Figure 67. Score card for table VI.

b. The exercise is fired from a stationary tank at moving targets (6 x 6-foot panels) at unknown ranges, varying from 700 to 1,500 yards. Either a powered target or a towed target may be used. To vary the range from tank to target, either the tank or the targets may be moved to different locations. The target will be exposed for approximately 300 yards and will travel that distance at a constant speed of between 8 and 15 miles per hour.

c. The examiner will lay the gun for direction for each target while issuing a six-element initial fire command. Time for each problem starts when the command FIRE is announced in the initial fire command.

d. The gunner will use the burst-on-target method of adjustment to fire the second round. This round will be fired even if the first round is a target hit.

e. Scoring will be conducted as indicated in table VII.

Table VII. Possible Score: 100 Points

Trial	Number rounds	Possible	Minus	Score
Target 1	2	20		
Target 2	2	20		
Target 3	2	20		
Target 4	2	20		
Target 5	2	20		

Cuts for each target:
    Failure to fire first round within 15 seconds_____ 5 points.
    Deduct 1 point for each additional second over
      15 to fire first round, up to 20 seconds_____ 5 points.
    Failure to hit target with first round_____ 5 points.
    Failure to hit target with second round_____ 5 points.

f. See figure 68 for illustration of an appropriate score card.

151. Table VIII: Range Card Firing Exercise

a. The purpose of this exercise is to test the ability of the gunner to determine prearranged firing data for selected targets and to engage area type targets successfully with HE ammunition under conditions of restricted visibility, and to afford night firing practice.

b. Five 6 x 6-foot panels will be placed in a wide lateral area, at ranges varying from 800 yards minimum to 3,500 yards maximum, and at different angles of site. Panels will be numbered consecutively from left to right and will be visible from a suitable firing position.

c. Examining personnel will accurately compute the following data for each panel:

    (1) The azimuth indicator reading from an aiming stake or reference point.

    (2) The quadrant elevation, gun to target (elevation for range plus angle of site), with the elevation quadrant. This is determined in the following manner:

        (*a*) Check and adjust the elevation quadrant with the M1 gunner's quadrant.

        (*b*) Determine the accurate range to the target.

        (*c*) Lay the aiming cross of the sight reticle on the center of the target, as would be done in direct fire.

        (*d*) Without disturbing the lay of the gun, measure the existing elevation with the elevation quadrant.

            Note. If the reading on the micrometer dial is between markings, record to the next higher whole mil.

Co **A**
Bn **1st**

NAME **SMITH, JOHN J.**
RANK **Pfc**   SN **38132977**
DATE **5 Feb 54**

POSSIBLE SCORE ___100___
TOTAL CUTS ___17___
TOTAL SCORE ___83___

TANK GUNNERY QUALIFICATION COURSE SCORE CARD

TABLE VII (SERVICE FIRING MOVING TARGET).

TRIAL	NUMBER OF ROUNDS	POSSIBLE POINTS	A TIME 1ST ROUND FIRED (Seconds)	B TARGET HIT 1ST ROUND		C TARGET HIT 2D ROUND	
				YES	NO	YES	NO
TARGET 1	2	20	11		X		X
2	2	20	12	X			X
3	2	20	14	X			X
4	2	20	17	X			X
5	2	20	13		X		X

Cuts:		MAXIMUM CUTS	TARGETS					TOTAL CUTS
			1	2	3	4	5	
A.	Failure to fire 1st round within 15 seconds	5 points	0	0	0	5	0	5
	(Deduct one point for each second over 15.) Maximum cut.	5 points	0	0	0	2	0	2
B.	Failure to hit target with 1st round	5 points	5	0	0	0	5	10
C.	Failure to hit target with 2d round	5 points	0	0	0	0	0	0

TOTAL **17**

Satisfactory Score _____ 70 points.

Lt Leonidas K. Jones
Examining Officer's Signature

Figure 68. Score card for table VII.

d. Ten E-type silhouette targets will be placed around the panel at which the gunner will fire. These panels will be placed as shown in figure 69.

e. The exercise will be conducted as follows:
(1) *Part I.* The gunner will be required to prepare a range card for the area, using the panels as likely targets. Information to be recorded on the range card will include—
(a) Aiming stake or reference point.
(b) Target (panel) numbers (left to right).
(c) Deflection (azimuth indicator reading) from aiming stake to each target.

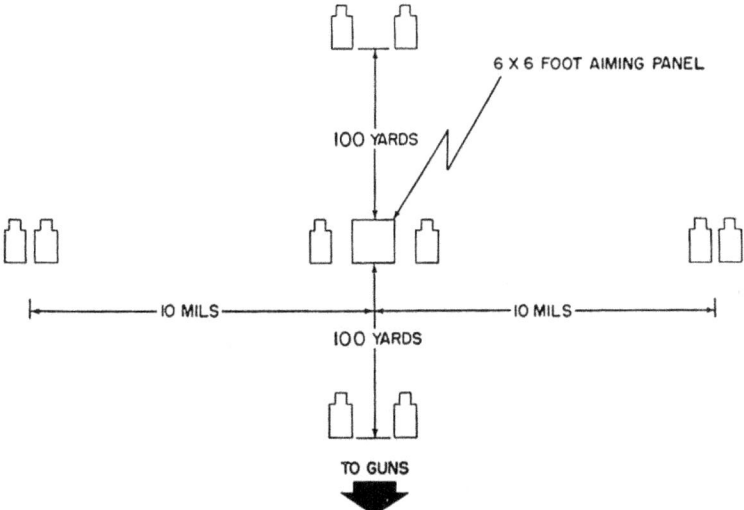

Figure 69. Range card firing exercise.

    (d) Range to each target in yards.
    (e) Quadrant for HE ammunition.
  (2) *Part II.* After the range card is prepared, the direct-fire sights will be covered (or the exercise may be fired at night) and the gunner will be required to fire on one of the panels, using his prearranged firing data. Using the data computed by the gunner to one of the targets, the examiner will issue an initial fire command. The gunner will set off the data as announced in the initial fire command and fire only the first round. The gunner will then add 1 mil in elevation and simulate firing the second round; drop 2 mils and simulate firing the third round; add 1 mil, traverse right 10 mils, and simulate firing the fourth round; traverse left 20 mils and simulate firing the fifth round. He will announce ON THE WAY as he simulates firing each round. Time will be recorded from the command FIRE.

    Note. Elevation will be changed 1 mil to effect a 100-yard range change.

f. The exercise will be scored as follows:

Table VIII. Possible Score: 100 Points

Part I_____Total value 50 points
    For failure to obtain correct azimuth indicator
      reading within plus or minus 1 mil on each target,
        cut 5 points for each mil (maximum cut 25 points)___25 points

> For failure to obtain correct quadrant reading within plus or minus 1 mil on each target, cut 5 points for each mil (maximum cut 25 points)_____25 points

Part II _____Total value 50 points

> For each five seconds, or fraction thereof, that the time to fire the first round exceeds 25 seconds (30 seconds if M1 gunner's quadrant is used), cut 5 points (maximum cut 25 points)_____25 points
>
> For failure to hit in target area (20 mils by 200 yards), cut_____25 points

g. See figure 70 for illustration of an appropriate score card.

Co __A__
Bn. __13T__

NAME __SMITH JOHN U.__
RANK __PFC__   SN __38132977__
DATE __6 FEB 54__

POSSIBLE SCORE __100__
TOTAL CUTS __15__
TOTAL SCORE __85__

TANK GUNNERY QUALIFICATION COURSE SCORE CARD

TABLE VIII (RANGE CARD FIRING EXERCISE)

PART I		ROUNDS					TOTAL CUTS
		1	2	3	4	5	
(50 points)	1. Failure to obtain correct azimuth indicator reading within plus or minus 1 mil on each target. (Cut 5 points for each mil error not to exceed a total of 25 points.)	5	0	0	5	0	10
	2. Failure to obtain correct quadrant reading within plus or minus 1 mil on each target (Cut 5 points for each mil error not to exceed a total of 25 points.)	0	5	0	0	0	5
PART II		TIME FIRST RD FIRED (SEC)					
(50 points)	1. Failure to fire first round within 25 seconds. (Cut 5 points for each five seconds or fraction thereof over 25 seconds not to exceed 25 points.)	19					0
	2. Failure to hit target area 20 mils by 200 yards. (Cut 25 points.)	HIT					0
						TOTAL CUTS	15

Satisfactory Score _____ 70 points.

Lt Leonidas K. James
Examining Officer's Signature

Figure 70. Score card for table VIII.

APPENDIX I
REFERENCES

DA Pam 108-1	Index of Army Motion Pictures, Television Recordings, and Filmstrips.
DA Pam 310-series	Military Publications.
AR 385-63	Regulations for Firing Ammunition for Training, Target Practice, and Combat.
AR 750-628	Maintenance of Supplies and Equipment.
SR 320-5-1	Dictionary of United States Army Terms.
SR 320-50-1	Authorized Abbreviations.
FM 5-25	Explosives and Demolitions.
FM 17-12	Tank Gunnery.
FM 21-5	Military Training.
FM 21-6	Techniques of Military Instruction.
FM 21-8	Military Training Aids.
FM 21-30	Military Symbols.
FM 21-60	Visual Signals.
FM 23-55	Browning Machineguns, Caliber .30, M1917A1, M1919A4, M1919A4E1, M1919A6, and M37.
FM 23-65	Browning Machineguns, Caliber .50, HB, M2.
TM 9-308A	76-mm Gun T91E3.
TM 9-730	76-mm Gun Tanks M41 (T41E1) and T41E2 (M41A1).
TM 9-1901	Artillery Ammunition.
TM 11-284	Radio Sets AN/GRC-3, -4, -5, -6, -7, and -8.
TM 11-704	Auxiliary Interphone Equipment, AN/VIA-1.

APPENDIX II

STOWAGE

Proper stowage of tank equipment is necessary for the efficient functioning of the tank and crew. First, each crew member must ascertain whether or not the equipment necessary to perform his duties is present. Second, and equally important, this equipment must be stowed in the proper place in order to be available when needed. In order that these conditions can be met, a list of vehicle stowage and special tools has been prepared for the vehicle. The list of vehicle stowage and special tools designates an exact location either on or within the tank for every piece of authorized equipment, including personal equipment.

Name of part	Quantity required per vehicle	Where carried
ARMAMENT:		
GUN, 76-mm, M32	1	In combination mount in turret.
MOUNTS, GUN:		
MOUNT, Cal .50 machinegun, turret mounted.	1	On turret.
EQUIPMENT FOR GUN, SUBMACHINE, CAL .45, M3A1:		
CASE, ammunition	1	In turret bustle.
EQUIPMENT FOR CARBINE, CAL .30, M2:		
CASE, ammunition	1	In turret bustle.
EQUIPMENT FOR GUN, MACHINE, CAL .30, BROWNING, M1919A4E1 (COAXIAL):		
BAG, assy, empty cartridge, cal .30.	1	On coaxial machinegun mount.
EQUIPMENT FOR GUN, MACHINE, CAL .50, BROWNING M2, HB (FLEX):		
COVER, cal .50 barrel	1	On spare barrel.
COVER, machinegun, cal .50	1	On machinegun.
COVER, tripod mount, M1	1	On tripod mount.
EQUIPMENT FOR GUN, 76-MM, M32:		
COVER, breech	1	On gun breech.
COVER, assy, muzzle	1	On gun.

Name of part	Quantity required per vehicle	Where carried
GUN, lubrication oil w/hose 2 oz.	1	In gun spare parts roll.
ROLL, gun spare parts and tools.	1	In turret bustle.
WRENCH, adjusting, housing spring.	1	In gun spare parts roll.
SPARE PARTS FOR GUN, 76-MM, M32, AND MOUNT, COMBINATION, M76:		
GASKET, recoil cylinder, filling plug.	2	In gun spare parts roll.
PLUG (filling recoil cylinder).	1	In gun spare parts roll.
PLUG, pipe, square head (recoil mechanism).	1	In gun spare parts roll.
PLUG, pipe square head, 1/8 in (replenisher).	1	In gun spare parts roll.
PLUG (replenisher)	1	In gun spare parts roll.
SIGHTING AND FIRE-CONTROL EQUIPMENT:		
MOUNT, Periscope, M93	1	In turret (for gunner's periscope).
MOUNT, Periscope, M94	1	In turret (for commander's periscope).
MOUNT, Telescope, M92 and M92A1.	1	In turret.
DRIVE, Ballistic, M4	1	In turret.
PERISCOPE, M20 (gunner's and commander's).	2	In turret.
SYSTEM, projection. Composed of:		
1—Mirror, assy, lower		In turret.
1—Mirror, assy, upper		In turret.
TELESCOPE, M97	1	In Mount, M92.
SPARES (ADDITIONAL SPARES LISTED W/SIGHTING MAJOR ITEMS):		
HEAD, assy, Periscope, M20 (gunner and commander).	2	In turret.
LAMP, electric, 3 v, No. 323 (for Ballistic Drive, M4)	4	In box.
VEHICULAR EQUIPMENT:		
BAG, pamphlet	1	In turret.
BOX, assembly spare bulbs	1	Turret bustle.
BOX, assy, signal flare complete.	1	Turret left wall.

AGO 4054B

Name of part	Quantity required per vehicle	Where carried
CABLE, towing, steel, diam 1⅛ in. length 10 ft, with 2 eyes, 1¼ x 3¼ in.	1	On left fender.
CORD, light ext, inspection, sgle-contact, plug and socket, length 15 ft.	1	Tool bag.
EXTENSION, lubr, gun type, hydraulic, 12 in. long.	1	Tool bag.
EXTINGUISHER, fire, portable 5 lb, CO_2.	1	Hull left wall.
GUN, lubr, hand lever operated, cap 15 oz.	1	In tool bag.
OILER, steel, pump type, cap 1 pt, 9 in spout.	1	Left fender box.
PADLOCK SET, 1¾ in. w/clevis, keyed interchangeably, composed of 4 locks and 6 keys.	1	Fender boxes and loader's hatch.
PAULIN, canvas, 12 x 12 ft.	1	On turret outside.
PUMP and hose assemblies, hand fuel (both elbows to be positioned to right of vertical centerline of pump).	1	In right fender box.
TAPE, adhesive, pressure-sensitive, water resistant, OD-7.4 in. wide, 15 yd lg.	1	In left fender box.
TAPE, friction, ¾ in. wide, 8 oz roll.	1	In left fender box.
TAPE, friction, ¾ in. wide 8 oz roll.	1	In tool bag.
WIRE, carbon, low annealed, soft, black, diam 0.080 in. (10 ft roll).	1	In tool bag.
ELBOW ASSEMBLY (for fixed fire extinguisher).	2	In tool bag.
COVER, azimuth indicator	1	On indicator.
BOLT, engine compartment door (combat locking).	11	In tool bag.
NUT, engine compartment door bolt, locking (self locking).	11	In tool bag.
LAMP, elec, 32 CP 24-28 v, MZ 1683.	1	In inspection cord light.
KIT, winterization, engine compartment covers.	1	On engine compartment grilles (when in use).

Name of part	Quantity required per vehicle	Where carried
PUBLICATIONS:		
FORM (envelope), DA 478	1	In pamphlet bag.
ORDER, Lubr, 9-730	1	In pamphlet bag.
CHART, strap location	1	In pamphlet bag.
SPARE PARTS, VEHICULAR:		
FITTING, lubrication straight, 1/8—27 NPT.	3	In tool bag.
LAMP, elec, No. 323 (for azimuth indicator).	3	In box.
LAMP, elec, 24-28 v, No. 313 (for panel lights).	3	In box.
TOOLS: COMMON AND TOOL EQUIPMENT:		
BAG, tool (w/o contents)	1	In right front fender box.
BAR, cross, socket wrench, round, 7/16 in. x 7 in., 9 in. long.	1	In tool bag.
BAR, jimmy, 1/2 in. x 11 7/8 in. long.	1	In turret.
BAR, socket wrench, extension, 1/2 in. sq dr 4 1/2–5 1/2 in. long.	1	In tool bag.
BAR, socket wrench, extension 1/2 in. sq dr 9 1/2–10 1/2 in. long.	1	In tool bag.
BAR, socket wrench, extension, 3/4 in. sq dr, 16 in. long.	1	In tool bag.
CHISEL, machinist's hand, cold, 3/4 in., 8 in. long.	1	In tool bag.
FILE, AS, hand, cut smooth, 10 in. long.	1	In tool bag.
FILE, AS, three-sq cut smooth, 6 in. long.	1	In tool bag.
HAMMER, machinist, ball peen, 2 lb.	1	In tool bag.
HANDLE, socket wrench, hinged, 1/2 in. sq dr, 16–18 in. long.	1	In tool bag.
HANDLE, socket wrench, ratchet, 1/2 in. sq dr, 9 1/2–15 in. long.	1	In tool bag.
HANDLE, socket wrench, speeder, 1/2 in. sq dr, 12 in. long.	1	In tool bag.

Name of part	Quantity required per vehicle	Where carried
HANDLE, socket wrench T-sliding, ½ in. sq dr, 9–11 in. long.	1	In tool bag.
HANDLE, socket wrench, T-sliding, ¾ in. sq dr, 17–19 in. long.	1	In tool bag.
JOINT, socket wrench, universal, ½ in. sq dr.	1	In tool bag.
PLIERS, lineman's side cutting, 8 in. long.	1	In tool bag.
PLIERS, combination, w/cutters, slip joint, 8 in. long.	1	In tool bag.
SCREWDRIVER, common, normal duty, single grip, 6 in. blade.	1	In tool bag.
SCREWDRIVER, common, normal duty, 1½ in. blade.	1	In tool bag.
SCREWDRIVER, mach extra heavy, 5 in. blade.	1	In tool bag.
WRENCH, adj sgle end, 12 in.	1	In tool bag.
WRENCH, adj, single end, 7½–8½ in.	1	In tool bag.
WRENCH, engrs, 15° angle, dble hd, OE, 5/16 in. and 3/8 in.	1	In tool bag.
WRENCH, engrs, 15° angle, dble hd, OE, 7/16 in. and ½ in.	1	In tool bag.
WRENCH, engrs, 15° angle, dble hd, OE, 9/16 in. and 11/16 in.	1	In tool bag.
WRENCH, engrs, 15° angle, dble hd, OE, 5/8 in. and ¾ in.	1	In tool bag.
WRENCH, engrs, 15° angle, dble hd, OE, 13/16 in. and 7/8 in.	1	In tool bag.
WRENCH, engrs, 15° angle, dble hd, OE, 15/16 in. and 1 in.	1	In tool bag.
WRENCH, engrs, single head, open end, 2¼ in.	1	In tool bag.
WRENCH, set or cap screw, hex, ⅛ in.	1	In tool bag.

Name of part	Quantity required per vehicle	Where carried
WRENCH, set or cap screw, hex, 5/32 in.	1	In tool bag.
WRENCH, set or cap screw, hex, 3/16 in.	1	In tool bag.
WRENCH, set or cap screw, hex, 1/4 in.	1	In tool bag.
WRENCH, set or cap screw, hex, 5/16 in.	1	In tool bag.
WRENCH, set or cap screw, hex, 3/8 in.	1	In tool bag.
WRENCH, set or cap screw, hex, 5/8 in.	1	In tool bag.
WRENCH, socket, 1/2 in. sq. dr, 12 pt, 7/16 in.	1	In tool bag.
WRENCH, socket, 1/2 in. sq. dr, 12 pt, 1/2 in.	1	In tool bag.
WRENCH, socket, 1/2 in. sq. dr, 12 pt, 9/16 in.	1	In tool bag.
WRENCH, socket, 1/2 in. sq. dr, 12 pt, 5/8 in.	1	In tool bag.
WRENCH, socket, 1/2 in. sq. dr, 8 pt, 3/8 in.	1	In tool bag.
WRENCH, socket, 1/2 in. sq. dr, 12 pt, 11/16 in.	1	In tool bag.
WRENCH, socket, 1/2 in. sq. dr, 12 pt, 3/4 in.	1	In tool bag.
WRENCH, socket, 1/2 in. sq. dr, 12 pt, 7/8 in.	1	In tool bag.
WRENCH, socket, 1/2 in. sq. dr, 12 pt, 15/16 in.	1	In tool bag.
WRENCH, socket, 1/2 in. sq. dr, 12 pt, 1 in.	1	In tool bag.
WRENCH, socket, 1/2 in. sq. dr, 12 pt, 1 1/16 in.	1	In tool bag.
WRENCH, socket, 1/2 in. sq. dr, 12 pt, 1 1/8 in.	1	In tool bag.
WRENCH, socket, 3/4 in. sq. dr, 12 pt, 1 3/16 in.	1	In tool bag.
WRENCH, socket, 3/4 in. sq. dr, 12 pt, 1 1/4 in.	1	In tool bag.
TOOLS, PIONEER:		
AXE, chopping, sgle bit, 4 lbs.	1	Right fender.
BAR, crow, pinch point, 5 ft long, 1 1/4 in. wide.	1	Hull rear outside.
HANDLE, mattock, 36 in. long.	1	Right fender.

AGO 4054B

Name of part	Quantity required per vehicle	Where carried
MATTOCK, pick, M1, w/o handle, 5 lb.	1	Right fender.
SHOVEL, general purpose, 0-handle, round point.	1	Right fender.
SLEDGE, blacksmith, dble face, 10 lb.	1	Turret rear outside.
ARMAMENT:		
GUN, Machine, Cal .30, Browning, M1919A4E1.	1	In coaxial mount in turret.
GUN, Machine, Cal .50, Browning M2, HB (Flexible).	1	On top of turret.
MOUNTS, GUN:		
MOUNT, Tripod, Machine gun, Cal .50, M3.	1	On cover of bustle box.
EQUIPMENT FOR GUN, MACHINE, CAL .30, BROWNING, M1919A4E1 (COAXIAL):		
BRUSH, cleaning, cal .30	4	In gun spare parts roll.
BRUSH, chamber cleaning, M6.	1	In gun spare parts roll.
CASE, cleaning rod, M1	1	In oddment tray.
COVER, spare barrel	1	On spare barrel.
EXTRACTOR, ruptured cartridge.	1	Oddment tray.
ROD, cleaning, jointed, M1	1	In Case, cleaning rod, M2.
WRENCH, combination, M6	1	In gun spare parts roll.
SPARE PARTS FOR GUN, MACHINE, CAL .30, BROWNING, M1919A4E1:		
PARTS, spare w/box, cal .30 machinegun, combat vehicle.	1	In box in oddment tray.
1—Bolt, assembly		
1—Box, spare parts (empty)		
1—Extractor, assembly		
1—Lever, cocking		
1—Pin, cocking lever		
1—Pin, firing assembly		
1—Rod, driving spring assembly.		
1—Sear		
1—Spring, driving		
1—Spring, sear, assembly		
1—Trigger		
BARREL, assy	1	In cover in turret.

Name of part	Quantity required per vehicle	Where carried
EQUIPMENT FOR GUN, MACHINE, CAL .50, BROWNING M2, HB (FLEX):		
BRUSH, cleaning, cal .50, M4.	4	In gun spare parts roll.
CASE, cleaning rod, M15	1	In rear turret stowage box.
EXTRACTOR, ruptured cartridge, cal .50.	1	Oddment tray.
GAGE, headspace and timing	1	In gun spare parts roll.
HIDER, flash, cal .50, HB	1	On cal .50 machine gun.
ROD, cleaning, jointed, cal .50, M7.	1	In Case, cleaning, M15.
WRENCH, combination, spanner and screwdriver.	1	In gun spare parts roll.
SPARE PARTS FOR GUN, MACHINE, CAL .50, BROWNING, M2, HB (FLEX):		
PARTS, spare, w/box, cal .50 machingun, combat vehicle consisting of: 1—Bolt, alternate feed assy 1—Box, spare parts (empty) 1—Extension, firing pin assy 1—Extractor, assy 1—Lever, cocking 1—Lock, accelerator stop 1—Pin, cocking lever 1—Pin, firing 1—Sear 1—Slide, sear 1—Spring, sear 1—Stop, accelerator 1—Switch, bolt	1	In box in oddment tray.
BARREL ASSEMBLY (spare).	1	In cover outside turret bustle.
EQUIPMENT FOR GUN, 76-MM, M32:		
BRUSH, bore, M15	2	Rear turret stowage box.
COVER, brush, bore, M516	2	On brush.
COVER, gun book, M539	1	In pamphlet bag.
EYEBOLT, breechblock removing.	1	In gun spare parts roll.
OIL, hydraulic, petroleum base (in 1 qt screw cap can).	1 qt.	Left rear fender box.

Name of part	Quantity required per vehicle	Where carried
RAMMER, cleaning and unloading, M3.	1	Rear turret stowage box.
ROD, spline (operating) shaft removing.	1	In oddment tray.
STAFF, middle, 48¼ in. long.	5	Left rear fender box.
RING, wiper	1	Rear turret stowage box.
TOOL, breechblock removing.	1	In gun spare parts roll.
TOOL, extracting and ramming.	1	In turret left side wall.
WRENCH, tubular pronged, single end.	1	In gun spare parts roll.
SPARE PARTS FOR GUN, 76-MM, M32, AND MOUNT, COMBINATION, M76:		
MECHANISM, percussion assy.	1	In gun spare parts roll.
SIGHTING AND FIRE-CONTROL EQUIPMENT:		
LIGHT, Instrument, M36 (for Telescope Mount, M92, and Periscope, M20).	3	
LIGHT, Instrument, M30 (for Ballistic Drive, M4).	1	In turret in clip.
PERISCOPE, M17	4	Around driver's hatch.
PERISCOPE, M13	1	In turret.
PERISCOPE, M19	1	In driver's compartment.
QUADRANT, Gunner's, M1, w/equipment, consisting of:		
CASE, Carrying, M82 OR	1	In turret or fender box.
CASE, Carrying, M18 OR		
CASE, Carrying, M56		
QUADRANT, Gunner's M1.	1	In case.
SETTER, fuse, M27 (wrench) OR	1	Roll, gun spare parts and tools.
SETTER, fuse M14 (wrench)	1	In pamphlet bag.
TABLE, firing, FT-76-C-1.		
SPARES (ADDITIONAL SPARES LISTED W/SIGHTING MAJOR ITEMS):		
HEAD, assy, Periscope, M19.	1	In driver's compartment.

Name of part	Quantity required per vehicle	Where carried
LAMP, elec, 3 v, No. 325 (for Instrument Lights, M36).	6	In box.
PERISCOPE, M17	1	Driver's compartment.
PERISCOPE, M13 (loader's).	1	Turret.
EQUIPMENT, MISCELLANEOUS:		
CAN, water, 5 gal standard	2	Side of turret stowage box.
CARRIER, wire-cutter, M1938.	1	In tool bag.
CUTTERS, wire, M1938	1	In tool bag in Carrier, wire cutter, M1938.
FLAG SET, M238, consisting of:	1	In right side of turret.
1—Case, CS-90		
1—Flag, MC-273 (red)		
1—Flag, MC-274 (orange).		
1—Flag, MC-275 (green).		
3—Flag, staff, MC-270		
FLASHLIGHT, elec, hand, 2 cell, w/lamp, w/o batteries.	3	One, hull left wall; two turret.
KIT, first aid, motor vehicle, 12 unit, size 1.	1	Turret bustle on left front fender box.
MITTENS, asbestos, M1942	2	In turret oddment tray.
PANEL SET, AP-50, consisting of:	1	Right rear fender box.
1—Panel 140, w/case		
1—Panel 141, w/case		
RADIO SETS AND COMBINATIONS:		
RADIO SET, AN/GRC-3 or AN/GRC-4, or AN/GRC-7 or AN/GRC-8, and Auxiliary Interphone Equipment AN/VIA-1.		Turret bustle.
OR		
RADIO SET, AN/GRC-3 or AN/GRC-7 and AN/ARC-3 and Auxiliary Interphone Equipment, AN/VIA-1.		

Name of part	Quantity required per vehicle	Where carried
STOVE, cooking, gasoline, M1942, 1-burner.	2	Rear turret stowage box.
TUBE, flexible nozzle, cam type (for refillable gasoline drums).	2	Rear turret stowage box.
PUBLICATIONS:		
MANUAL, Technical, 9-730	1	In pamphlet bag.
AMMUNITION:		
ROUNDS, for Carbine, Cal .30 M2 (in. 30-rd magazines).	90	In case.
ROUNDS for Submachine-gun, Cal .45, M3A1 (in. 30-rd magazines).	180	In case.
ROUNDS for Gun, Machine, Cal .30 (in link metallic belts) (for coaxial gun).	5,000	750 in turret magazine, 4250 (in standard boxes) in turret.
ROUNDS for Gun, Machine, Cal .50 (AA) (in link metallic belts in 6 ammo boxes).	630	Rear turret stowage box.
ROUNDS for Gun, 76-mm, M32, M41.	57	33 in hull, 24 in turret.
SUGGESTED BASIC LOAD BY TYPE:		
HE		36 rounds
AP		10 rounds
HVAP-DS		5 rounds
WP		6 rounds
ROUNDS for Gun, 76-mm, M32, M41A1.	65	33 in hull, 32 in turret.
SUGGESTED BASIC LOAD BY TYPE:		
HE		40 rounds
AP		12 rounds
HVAP-DS		6 rounds
WP		7 rounds
GRENADES, hand		
Fragmentation, w/fuse	4	In turret.
Smoke (WP), w/fuse	2	In turret.
Incendiary, w/fuse	2	In turret.
ARMAMENT:		
GUN, Submachine, Cal .45, M3A1, w/sling.	1	In turret bustle.
CARBINE, Cal .30, M2, w/sling and oiler.	1	In turret.

Name of part	Quantity required per vehicle	Where carried
SIGHTING AND FIRE-CONTROL EQUIPMENT: BINOCULAR, M17A1, w/equipment, consisting of— 　1—Binocular, M17A1 　1—Case, Carrying	1	In Case, M63A1.
EQUIPMENT, MISCELLANEOUS: BATTERY, Flashlight, BA30.	14	Six installed in flashlights; eight installed in instrument lights.
CANTEEN, M1910, complete w/cup and cover.	4	One on hull left wall, three in turret.
PACK, field, cargo and combat, M1945, consisting of— 　1—Pack, field, cargo, M1945. 　1—Pack, field, combat, M1945. 　1—Suspenders, pack, field, cargo and combat.	4	Outside turret.
RATIONS, Individual combat.	12	Four in turret bustle, eight in fender boxes.

INDEX

	Paragraph	Page
Abandon tank	83	90
Accumulator hand pump and handle	23	36
Action:		
Dismounted	82–84	90
Mounted	76–81	83
Adjustment:		
Tests and, gunner's quadrant	42	60
Using the azimuth indicator and the elevation quadrant, test.	141	136
Direct fire, primary method	102, 110	109, 118
Fire, alternate method	103, 110	109, 118
Solenoid, coaxial Cal .30 MG	15	29
After-firing checks	13	27
After-operation service	96	101
Alternate method, adjustment of fire:		
Moving targets	110	118
Stationary targets	103	109
Ammunition:		
Required, tank gunnery qualification course	125	129
Stowage and handling	81	89
Test on identification and inspection	136	133
Types	6	9
Assembly:		
Recoil cylinder	11	23
Replenisher	11	23
Armament	4	7
At-the-halt service	95	99
Auxiliary fire control:		
Equipment	35	48
Exercise	146	144
Azimuth indicator, M31	41	58
Test on adjustment using the elevation quadrant and	141	136
Backlash	51	71
Ballistic drive, M4	38	53
Test on direct laying, primary sighting devices	137	133
Test on direct laying, secondary sighting devices	138	134
Battle sight	116	124
Before-firing checks	13	27
Before-operation service	93	96
Blast deflector (muzzle brake)	5	8
After-firing care	13	27
Bleeding, recoil system	11	23
Block, vision	47	67
Bore evacuator	5	8
After-firing care	13	27

	Paragraph	Page
Boresighting	50	68
Coaxial Cal .30 MG	15	29
Box:		
Gun control, M41A1	31	44
Turret control, M41	24	37
Breech:		
After-firing care	13	27
Components	7	10
Manual opening	10	21
Mechanism	5, 7	8, 10
Operation	7	10
To open	71	82
Breechblock:		
Test on assembly	130	132
Test on disassembly	129	131
Card, range	120	126
Firing exercise	151	155
Casualty:		
To evacuate from driving compartment	88	92
To evacuate from turret	89	93
Characteristics	2	4
Description, M41 and M41A1	3	4
Checking replenisher assembly	11	23
Classification of gunners	124	129
Clear and secure guns	79	88
Close and open hatches drill	64	78
Coaxial machinegun:		
Firing	112	122
Mount	15	29
Cocking	10	21
And firing mechanisms, Gun 76-mm, M32	8	15
Combination gun mounts M76 and M76A1	15	29
Commander's power control:		
M41	21	34
M41A1	29	42
Commands:		
Fire:		
Moving targets:		
Initial	108	115
Subsequent	111	120
Stationary targets:		
Initial	100	105
Subsequent	104	112
For control of turret	61	75
For movement of tank	61	75
Conduct of fire	98	104
Control:		
Box:		
Gun, M41A1	31	44
Turret, M41	24	37

	Paragraph	Page
Control—Continued		
Commander's power:		
M41	21	34
M41A1	29	42
Firing:		
M41	25	38
M41A1	33	46
Gunner's manual:		
M41	23	36
M41A1	30	43
Gunner's power:		
M41	20	34
M41A1	28	42
Crew:		
Composition	53	72
Control:		
Box positions	56	73
Checking interphone equipment	59	74
Interphone language	61	75
Operation of interphone and radio	55	72
Radio check	58	74
Use of definite terminology	60	75
Drill:		
And service of the piece	52	72
Dismounted	62	76
Pep	67	80
To close and open hatches	64	78
To dismount tank	65	78
To dismount through escape hatch	66	79
To mount the tank	63	77
Gun positions, mounted	69	80
Preventive maintenance	91	95
Cylinder assembly, recoil	11	23
Data	4	7
Defilade, firing	121	127
Definite terminology, use	60	75
Destroy tank, action	84	91
Destruction of equipment	90	94
Direct-fire equipment	35	48
Direct laying:		
Test on primary sighting devices	137	133
Test on secondary sighting devices	138	134
Disassembly and assembly M32 76-mm gun	9	17
Dismount:		
Tank crew drill	65	78
Through escape hatch	66	79
Dismounted:		
Actions	82, 83, 84	90, 91
Drill	62	76
Posts	54	72
Doubtful sensing	101	108
Draining replenisher assembly	11	23

	Paragraph	Page
Drill:		
Crew, see crew		
Evacuation, general	87	92
Drive, ballistic, M4	38	53
During-firing checks	13	27
During-operation service	94	98
Duties:		
Firing	99	104
In firing or gun drill	78	87
Elements, fire commands:		
Initial	100, 108	105, 115
Subsequent	104, 111	112, 120
Elevation quadrant, M9	43	62
Test on adjusting	135	133
Test on adjustment using the azimuth indicator and the	141	136
Test on use	139	135
Equipment:		
Destruction	90	94
Sighting and fire-control, arrangement and use	35	48
Evacuation:		
Drill, general	87	92
Methods	86	91
Of wounded from tanks	85	91
Examination, gunner's preliminary	127	130
Exercises:		
Service firing, see Service firing exercises		
Subcaliber firing, see Subcaliber firing exercises		
External interphone	57	73
Extract and eject, failure	12	26
Extraction and ejection, 76-mm gun, M32	10	21
Filling replenisher assembly	11	23
Fire:		
Auxiliary control exercise	146	144
Commands:		
Moving targets:		
Initial	108	115
Subsequent	111	120
Stationary targets:		
Initial	100	105
Subsequent	104	112
Conduct	98	104
Control and sighting equipment:		
Azimuth indicator, M31	41	58
Ballistic drive, M4	38	53
Boresighting and zeroing	50	68
Elevating Quadrant, M9	43	62
Fuze Setter, M27	49	68
Gunner's Quadrant, M1 and M1A1	42	60
Periscope:		
M13	45	65
M17	44	64
M19	46	67

	Paragraph	Page
Fire—Continued		
Control of sighting equipment—Continued		
Periscope—Continued		
M20	36	49
Mounts, M93 and M94	37	51
Projection system	48	67
Synchronization and backlash	51	71
Telescope, M97	39	53
Mounts, M92 and M92A1	40	56
Vision blocks	47	67
Failure	12	26
Massed	118	125
Ricochet	117	125
Firing:		
At moving targets:		
Adjustment of fire	110	118
General	105	114
Initial fire commands and firing duties	108	115
Leading	106	114
Sensings	109	118
Subsequent fire commands	111	120
Tracking	107	114
At stationary targets:		
Adjustment of direct fire, primary method	102	109
Adjustment of fire, alternate method	103	109
Initial fire commands and firing duties	100	105
Sensings	101	108
Subsequent fire commands	104	112
Cal .50 MG	16, 112	32, 122
Coaxial Cal .30 MG	15, 112	29, 122
Duties	99	104
Exercises:		
Service. (*see* Service exercises)		
Subcaliber. (*see* Subcaliber exercises)		
From defilade	121	127
Gun, 76-mm, M32	10	21
Night	119	126
Premature	12	26
Fleeing targets	115	123
Functioning, 76-mm gun, M32	10	21
Fuze setter, M27	49	68
Gun:		
Control box, M41A1	31	44
Laying reticle	36, 39	49, 53
To load	72	82
76-mm, M32:		
Breech mechanism	7	10
Care, cleaning, and lubrication	13	27
Cocking and firing mechanisms	8	15
Data	6	9
Disasembly and assembly	9	17
Functioning	10	21

	Paragraph	Page
Gun—Continued		
76-mm, M32:—Continued		
General	5	8
Malfunctions	12	26
Recoil mechanism	11	23
Gunner('s):		
Classification	124	129
Manual control:		
M41	23	36
M41A1	30	43
Power control:		
M41	20	34
M41A1	28	42
Preliminary examination, general	127	130
Quadrant, M1 (M1A1)	42	60
Test on testing	134	133
Test on use	140	136
Gunnery qualification course, tank	122	128
Handling and stowage of ammunition	81	89
Hatches:		
Dismount through escape	66	79
Open and close (drill)	64	78
Indicator azimuth, M31	41	58
Test on adjustment using the elevation quadrant and	141	136
Initial fire commands and firing duties:		
Moving targets	108	115
Stationary targets	100	105
Installation:		
Cal .50 MG	16	32
Coaxial Cal .30 MG	15	29
Instrument light	36	49
Interphone:		
Checking equipment	59	74
Language	61	75
Operation	55, 57	72, 73
Use of definite terminology	60	75
Language, interphone	61	75
Leading	106	114
Load:		
Failure	12	26
Weapons	80	89
Loader's traverse safety, M41	22	34
Loading:		
Cal .30 MG	15	29
Gun, 76-mm, M32	10	21
Local security, dismounted	82	90
Lock:		
Gun traveling	26	40
Turret	26	40
Lost sensing	101	108

	Paragraph	Page
Machineguns:		
Cal .50, M2, HB	16	32
Coaxial Cal .30	15	29
Firing	112	122
Maintenance:		
Crew preventive	91	95
Crew procedures:		
After-operation service	96	101
At-the-halt service	95	99
Before-operation service	93	96
During-operation service	94	98
Weekly preventive service	97	102
Test on care and	131	132
To be performed	92	95
Manipulation exercises, 1,000-inch	143	137
Massed fire	118	125
Materiel tasts:		
Test on adjusting elevation quadrant	135	133
Test on assembly of breechblock	130	132
Test on care and maintenance	131	132
Test on disassembly of breechblock	129	131
Test on identification and inspection of ammunition	136	133
Test on putting the turret into power operation	133	133
Test on sight adjustment	132	132
Test on testing gunner's quadrant	134	133
Methods of evacuation	86	91
Misfire	73	82
Modes of operation, interphone, radio	57	73
Monitoring	57	73
Mounted:		
Actions	76–81	83
Posts	54	72
Mounts:		
Caliber .50 MG, turret	16	32
Combination, M76 and M76A1	15	29
General, machine gun	14	29
Maintenance of machine gun	18	33
Tripod, M3	17	33
Mount the tank crew, drill	63	77
Movement of the tank, commands	61	75
Moving:		
Tank exercise	145	141
Target exercise (shot)	150	153
Moving targets:		
Adjustment of fire	110	118
General	105	114
Initial fire commands and firing duties	108	115
Leading	106	114
Sensings	109	118
Subsequent fire commands	111	120
Tracking	107	114
Night firing	119	126

	Paragraph	Page
Open and close hatches drill	64	78
Operation:		
Interphone and radio	55	72
Modes of interphone and sets 1 and 2	57	73
Over sensing	101	108
Pep drill	67	80
Periscope:		
Mount, M93 and M94	37	51
M13	45	65
M17	44	64
M19	46	67
M20	36	49
Boresighting	50	68
Test on direct laying, primary sighting devices	137	133
Possible scores, table of, gunner's preliminary examination	128	131
Posts, dismounted and mounted	54	72
Power operation:		
Steps for placing turret into	34	47
Test for putting the turret into	133	133
Precautions, safety	70	80
Preliminary examination, gunner's	127	130
Premature firing	12	26
Prepare to fire	77	83
Preventive maintenance:		
Crew	91	95
Weekly service	97	102
Primary method, adjustment of fire:		
Moving targets	110	118
Stationary targets	102	109
Procedure:		
To evacuate casualty from driving compartment	88	92
To evacuate casualty from turret	89	93
Projectile, to remove a stuck	75	83
Projection system	48	67
Purpose and scope:		
Manual	1	3
Tank gunnery qualification course	123	128
Quadrant:		
Elevation, M9	43	62
Gunner's M1	42	60
Test on adjusting elevation quadrant	135	133
Test on testing gunner's quadrant	134	133
Test on use of elevation quadrant	139	135
Test on use of gunner's quadrant, M1	140	136
Qualification course, tank gunnery	122	128
Radio:		
Check	58	74
Operation	55	72
Operation of sets 1 and 2	57	73
Range card	120	126
Firing exercise	151	155

	Paragraph	Page
Recoil cylinder assembly	11	23
Recoil mechanism, gun, 76-mm, M32	11	23
Record firing, rules	126	130
References	App I	159
Removal:		
Cal .50 MG	16	32
Coaxial Cal .30 MG	15	29
Stuck projectile	75	83
Stuck round	74	82
Unfired round or misfire	73	82
Replenisher assembly	11	23
Reticle:		
Gun-laying, M20 periscope	36	49
Gun-laying, M97 telescope	39	53
Ricochet fire	117	125
Round:		
Stuck	74	82
Unfired	73	82
Rules for record firing	126	130
Safety:		
Loader's traverse, M41	22	34
Manual	25	38
Precautions	70	80
Timing relay	25, 33	38, 46
Scope, purpose:		
Manual	1	3
Tank gunnery qualification course	123	128
Scores, table of possible, gunner's preliminary examination	128	131
Secure and clear guns	79	88
Security, local, dismounted	82	90
Sensings:		
Moving targets	109	118
Stationary targets	101	108
Service:		
After-operation	96	101
At-the-halt	95	99
Before-operation	94	98
During-operation	93	96
Of the piece	68	80
Weekly preventive maintenance	97	102
Service firing exercise	147	149
Table V	148	149
Table VI	149	152
Table VII	150	153
Table VIII	151	155
Setter, fuze, M27	49	68
Short sensing	101	108
Shot adjustment, moving target exercise, 200 feet	144	138
Sight:		
Battle	116	124
Test on adjustment	132	132

	Paragraph	Page
Simulated firing tests:		
Test on:		
Adjustment using the azimuth indicator and the elevation quadrant.	141	136
Direct laying, primary sighting devices	137	133
Direct laying, secondary sighting devices	138	134
Use of elevation quadrant	139	135
Use of gunner's quadrant, M1	140	136
Situations, special. (*See* Special situations).		
Solenoid, adjustment of Cal .30 MG	15	29
Special situations	113	123
Battle sight	116	124
Dangerous surprise targets	114	123
Firing from defilade	121	127
Fleeing targets	115	123
Massed fire	118	125
Night firing	119	126
Range cards	120	126
Ricochet fire	117	125
Stationary targets:		
Adjustment of direct fire, primary method	102	109
Adjustment of fire, alternate method	103	109
Initial fire commands and firing duties	100	105
Sensings	101	108
Subsequent fire commands	104	112
Stowage	App II	160
And handling of ammunition	81	89
Stuck:		
Projectile	75	83
Round	74	82
Subcaliber firing exercises	142	137
Table I	143	137
Table II	144	138
Table III	145	141
Table IV	146	144
Subsequent fire commands:		
Moving targets	111	120
Stationary targets	104	112
Switches:		
Dump valve toggle	23	36
Turret motor, M41A1	32	45
Synchronization	51	71
Table:		
Of possible scores, gunner's preliminary exam	128	131
Service firing	147	149
V—Zeroing exercise	148	149
VI—Exercise, stationary targets at variable ranges (shot and HE adjustment).	149	152
VII—Moving target exercise (shot)	150	153
VIII—Range card firing exercise	151	155

	Paragraph	Page
Table—Continued		
Subcaliber firing:		
I—Manipulation exercise, 1,000-inch	143	137
II—Shot adjustment, moving target exercise, 200 feet.	144	138
III—Moving tank exercise	145	141
IV—Auxiliary fire-control exercise	146	144
Tank:		
Commands for movement	61	75
Data, M41 and M41A1	4	7
Evacuation of wounded	85	91
Gunnery qualification course	122	128
Moving, exercise	145	141
To abandon	83	90
To destroy	84	91
Targets:		
Dangerous surprise	114	123
Fleeing	115	123
Moving:		
Adjustment of fire	110	118
General	105	114
Initial fire commands and firing duties	108	115
Leading	106	114
Sensings	109	118
Subsequent fire commands	111	120
Tracking	107	114
Stationary:		
Adjustment of direct fire, primary method	102	109
Adjustment of fire, alternate method	103	109
Initial fire commands and firing duties	100	105
Sensings	101	108
Subsequent fire commands	104	112
Telescope:		
Boresighting, M97	50	68
Mounts, M92 and M92A1	40	56
M97	39	53
Test on direct laying, secondary sighting devices	138	134
Terms	61	75
Test on:		
Adjusting elevation quadrant	135	133
Assembly of breechblock	130	132
Care and maintenance	131	132
Disassembly of breechblock	129	131
Identification and inspection of ammunition	136	133
Putting the turret into power operation	133	133
Sight adjustment	132	132
Testing gunner's quadrant	134	133
Tests:		
Materiel, see Materiel tests		
Simulated firing, see Simulated firing tests		
Tracking	107	114
Traveling lock, gun	26	40

	Paragraph	Page
Tube	5	8
After firing care	13	27
Turret:		
And armament controls and equipment:		
M41	19	33
M41A1	27	42
Commands for control	61	75
Control box	24	37
Lock	26	40
Motor switch	32	45
Placing into power operation	34	47
Unfired round	73	82
Unloading:		
Stuck:		
Projectile	75	83
Round	74	82
Unfired round or misfire	73	82
Use:		
Definite terminology	60	75
Test on, elevation quadrant	139	135
Test on, gunner's quadrant	140	136
Vision blocks	47	67
Weapons, to load	80	89
Zeroing	50	68
Exercise, service firing	148	149

[AG 470.8 (4 Nov 55)]

By Order of *Wilber M. Brucker*, Secretary of the Army:

MAXWELL D. TAYLOR,
General, United States Army,
Chief of Staff.

OFFICIAL:
JOHN A. KLEIN,
Major General, United States Army,
The Adjutant General.

DISTRIBUTION:

Active Army:
CNGB (2)
Tec Svc DA (1)
Tec Svc Bd (2)
Hq CONARC (18)
OS Maj Comd (5)
OS Base Comd (2)
Log Comd (2)
MDW (1)
Armies (10)
Corps (5)
USMA (2)
Inf Sch (25)
Armd Sch (50)
PMST ROTC Units (2)
Mil Dist (2)
Units organized under
  following TOE:
    7-26R, Hq,H&S Co,
    Armd Inf Bn (2)

17-1R, Hq, Armd Div (6)
17-2R, Hq Co, Armd Div (2)
17-17R, Tk Co, 76-mm Gun, Sep,
  (Sp) (5)
17-22R, Hq&HqCo, CC, Armd Div (2)
17-26R, Hq, H&S Co, Tk Bn, 90-mm
  Gun (2)
17-32R, Hq&HqCo, Armd Gp (2)
17-36R, Hq, H&S Co, Tk Bn, 120-mm
  Gun (2)
17-46R, Hq, H&S Co, Recon Bn, Armd
  Div (2)
17-52R, Hq&HqCo, Armd Cav Regt
  (2)
17-56R, Hq&HqCo, Armd Cav Recon
  Bn (2)
17-57R, Recon Co (4)
17-62R, Hq&Hq Co, Armd Div Tn (1)

NG: State AG (6); units—same as Active Army except allowance is one copy to each unit.

USAR: Same as Active Army except allowance is one copy to each unit.
For explanation of abbreviations used, see SR 320-50-1.

©2013 Periscope Film LLC
All Rights Reserved
ISBN#978-1-940453-08-8
www.PeriscopeFilm.com

www.ingramcontent.com/pod-product-compliance
Lightning Source LLC
Chambersburg PA
CBHW071710090426
42738CB00009B/1726